Principles, Applications, and Advances of Organic Reaction Mechanisms

Rabab M. Nasser
Egyptian Petroleum Research Institute, Egypt

IGI Global
Scientific Publishing
Publishing Tomorrow's Research Today

Vice President of Editorial	Melissa Wagner
Director of Acquisitions	Mikaela Felty
Director of Book Development	Jocelynn Hessler
Production Manager	Mike Brehm
Cover Design	Jose Rosado

Published in the United States of America by
IGI Global Scientific Publishing
701 East Chocolate Avenue
Hershey, PA, 17033, USA
Tel: 717-533-8845
Fax: 717-533-7115
Website: https://www.igi-global.com E-mail: cust@igi-global.com

Library of Congress Cataloging-in-Publication Data

Names: Nasser, Rabab, 1983- editor
Title: Principles, applications, and advances of organic reaction mechanisms / edited by Rabab Nasser.
Description: Hershey, PA : IGI Global Scientific Publishing, [2025] | Includes bibliographical references and index. | Summary: "The primary objective of this book is to offer a clear and concise understanding of organic reaction mechanisms. It aims to present a systematic and logical approach to elucidating reaction mechanisms"-- Provided by publisher.
Identifiers: LCCN 2024053222 (print) | LCCN 2024053223 (ebook) | ISBN 9798369364734 hardcover | ISBN 9798369364741 paperback | ISBN 9798369364758 ebook
Subjects: LCSH: Organic reaction mechanisms
Classification: LCC QD502.5 .P75 2025 (print) | LCC QD502.5 (ebook) | DDC 547/.2--dc23/eng/20250415
LC record available at https://lccn.loc.gov/2024053222
LC ebook record available at https://lccn.loc.gov/2024053223

British Cataloguing in Publication Data
A Cataloguing in Publication record for this book is available from the British Library.

Table of Contents

Detailed Table of Contents

Understanding the fundamental principles of chemical reactivity is crucial in the field of organic chemistry. This chapter explores three key concepts that govern the reactivity of organic compounds: acids, bases, and nucleophiles. The Brønsted-Lowry theory is introduced, which defines acids as proton donors and bases as proton acceptors. The strength of acids and bases, as measured by their pKa and pKb values, is discussed in detail, highlighting how it determines their ability to participate in proton-transfer reactions. Factors influencing acid and base strength, such as electronegativity, resonances, and inductive effects, were examined. The chapter also covers the role of nucleophiles in organic transformations, emphasizing how their reactivity depends on factors like charge, polarizability, and steric effects. Understanding the interplay between acids, bases, and nucleophiles provides a foundation for predicting and controlling the outcomes of a wide range of organic reactions, enabling the rational design of synthetic strategies and the development of new chemical processes.

Substitution and elimination are considered the most important types of reactions in organic chemistry. Substitution reactions are chemical reactions that involve the replacement of one functional group in a molecule or ion with another functional group. During the substitution, the bond between the functional group and the reactive centre is broken, and a new bond forms between that centre and the new functional group. On the other hand, elimination reactions are a type of organic chemical reaction in which a pair of atoms or groups are removed from a molecule. It is the primary process by which organic compounds with only one carbon-carbon bond (saturated compounds) are converted to compounds with two or three carbon-carbon bonds (unsaturated compounds). It is the primary process by which organic compounds with only one carbon-carbon bond (saturated compounds) are converted to compounds with two or three carbon-carbon bonds (unsaturated compounds).

The coupled transfer of electrons and protons underlies a broad range of energy conversion processes. Proton-coupled electron transfer (PCET) reactions play an essential role in photosynthesis and respiration, as well as in many electrochemical reactions. Thus, understanding the fundamental theoretical principles of PCET reactions is critical for the design based on electrochemical reaction. A dipole-electric field coupling is quite reasonable as a model that captures the major effect of a polar solvent on the transferring proton. To investigate the analog of the solvent surfaces for electron transfer shown, we may use a four-state basis, where the base functions consist of electron and proton initial and final states (localized in the respective reactant and product wells for both the electron and the proton). This chapter elaborates the role of PCET reaction in human body like a proton pump and several other biogenic enzymes and general (adiabatic) synthetic reactions of mechanism and applications of medicinal chemistry.

Reactive intermediates are chemical species that are short lived and unstable molecules formed during the overall reaction pathway. These short-lived intermediates are formed via homolytic or heterolytic cleavage of bond. The transient nature of intermediates makes them special. During the reaction they form and are nearly instantly consumed, typically in a matter of seconds. Reactive intermediates are essential for comprehending the reaction mechanism. They are effective catalysts for chemical change because of their instability; they "want" to transform into more stable forms and hence propel the reaction to its conclusion. These transitory entities mediate a series of chemical reactions necessary for the production of dyes and dye intermediates. They offer the key steps in the synthetic pathways that enable us to efficiently and more sustainably manufacture complex compounds.

Spectroscopic methods represent key tools for identifying organic reaction mechanisms and obtaining information about molecular structures and transient intermediates. However, the broad repertoire of organic reactions, frequent difficulties in tracking compound intermediates, and challenges in the interpretation of spectral data can hamper progress and lead to misconceptions. This chapter aims to explain these methods alongside kinetic studies and isotope effects to comprehensively understand reaction dynamics. This chapter explains how nuclear magnetic resonance, infrared spectroscopy, ultraviolet–visible spectroscopy, and computer-based mass spectrometry reveal interactions and highlights the importance of such approaches through comparison and spectral simulations. The finalized section demonstrates the need for a variety of perspectives and overall explanations for promoting the synthesis of complex organic compounds. The field looks forward to significant progress with further real-time perspectives and education on reliable machine learning using spectroscopy.

Sulfur imidations can be addressed by several synthetic approaches, depending on nature of the starting sulfide or sulfoxide. A wide range of imidating agents has been developed including N-haloamides and related reagents, azides, heterocyclic nitrene precursors, and iodinanes. All are characterised by their specific properties and reactivities. Most of the imidation procedures have been developed for sulfide imidation, while only a few are efficient for sulfoxide conversions. This is particularly true for metal-free variants of imidation reactions. Importantly, the nature of nitrogen-protecting group. This chapter demonstrates synthesis of sulfonamide using amines and thiol through different oxidative and non-oxidative rearrangement reactions. The significance of sulfoxides as commodity chemicals, their efficacy as versatile building blocks for the assembly of chemically and biologically useful molecules, including therapeutic agents, is increasing enantiopure-sulfoxides, owing to their important biological activities are highly useful in pharmaceutical industry.

Preface

The fascinating world of organic chemistry is defined by its ability to transform simple compounds into a vast array of complex molecules, each possessing unique structures and properties. At the core of this transformative process lies the study of organic reaction mechanisms—an essential concept that underpins the synthesis of new molecules. As the field continues to evolve, it is crucial to understand the principles that govern these transformations, and how the mechanisms behind them enable the creation of materials with diverse functionalities.

It is with great pleasure that I present *Principles, Applications, and Advances of Organic Reaction Mechanisms*, a reference work that aims to provide a thorough and insightful examination of the principles, applications, and advances in this ever-growing area of organic chemistry. This book is the culmination of contributions from experts across the field, and its goal is to serve as a comprehensive resource for a broad range of readers—from students and educators to professionals and researchers in academia and industry.

The primary aim of this book is to elucidate the intricate and often complex processes that govern organic reactions. Through a systematic and logical approach, it explores the factors influencing reaction pathways—such as electronic effects, steric hindrance, and solvent effects—and delves into the practical applications of organic reaction mechanisms in synthesizing molecules. Furthermore, this work highlights recent breakthroughs and emerging trends, including advancements in catalysis, green chemistry, and computational techniques, which are shaping the future of organic chemistry.

Whether you are an undergraduate student seeking to build a strong foundation in reaction mechanisms, a graduate student or researcher exploring advanced topics, or an educator or industrial professional seeking a reliable resource, this book will guide you through the essential concepts and cutting-edge developments in organic reaction mechanisms.

It is my hope that this book will inspire further exploration and innovation in the field of organic chemistry and that it will be an invaluable tool for those who wish to deepen their understanding of the mechanisms that form the backbone of organic transformations.

Organization of the Book

This volume begins with Chapter 1, which lays a critical foundation by examining the fundamental concepts of chemical reactivity through the lens of acids, bases, and nucleophiles. It introduces key theoretical frameworks such as the Brønsted-Lowry acid-base theory and elaborates on parameters like pKa and pKb values to explain reactivity trends. By exploring how factors such as electronegativity, resonance, and steric hindrance shape chemical behavior, the chapter equips readers with essential tools for predicting and rationalizing organic transformations. The role of nucleophiles, in particular, is discussed in depth, establishing the groundwork for understanding a vast array of mechanistic pathways.

Chapter 2 delves into the mechanisms of substitution and elimination reactions, two of the most pivotal transformations in organic chemistry. By dissecting the processes by which functional groups are either replaced or removed to generate unsaturated compounds, this chapter underscores the mechanistic diversity and synthetic utility of these reactions. It explores how these processes govern the structural modification of molecules and prepare students and researchers to engage with more complex synthetic strategies.

In Chapter 3, the focus shifts to proton-coupled electron transfer (PCET), an increasingly significant area in both fundamental and applied chemistry. This chapter bridges concepts from physical chemistry and biochemistry to describe the dual movement of electrons and protons in critical processes such as respiration, photosynthesis, and enzymatic catalysis. Notably, it connects PCET to medicinal chemistry applications and the behavior of bio-relevant molecules, offering both theoretical insights and practical relevance in drug discovery and synthesis.

Chapter 4 introduces the concept of reactive intermediates—fleeting species such as carbocations, carbanions, free radicals, and carbenes that form the heart of many reaction mechanisms. Their transient nature and high reactivity make them central to understanding reaction pathways and rates. This chapter illuminates how these intermediates are generated, stabilized, and harnessed to construct complex organic architectures, offering indispensable knowledge for anyone involved in synthetic design or mechanistic investigation.

Chapter 5 highlights the indispensable role of spectroscopic methods in elucidating reaction mechanisms. With a focus on tools such as NMR, IR, UV-Vis, and mass spectrometry, it provides insight into how structural and kinetic data can reveal intricate details about chemical transformations. The discussion extends to isotope effects and kinetic studies, as well as emerging techniques involving machine learning for spectral interpretation, thus equipping readers with modern approaches to mechanistic analysis.

The synthetic potential of sulfur-containing compounds is explored in Chapter 6, which presents innovative rearrangement-based strategies for synthesizing sulfoxides and related molecules. It examines both oxidative and non-oxidative pathways and their implications in drug design and industrial chemistry. Highlighting the growing pharmaceutical interest in enantiopure sulfoxides and sulfonamides, this chapter showcases how mechanism-driven approaches can enhance the precision and efficiency of therapeutic molecule development.

Finally, Chapter 7 investigates the diverse world of isomerism, a phenomenon that dramatically influences molecular behavior and function. From constitutional isomers to stereoisomers—including geometric and optical variants—this chapter dissects the structural subtleties that can alter a compound's physical, chemical, and biological properties. It emphasizes the practical implications of isomerism in areas like drug efficacy and molecular recognition, reaffirming the centrality of stereochemical understanding in both academic and industrial settings.

Together, these chapters offer a richly detailed journey through the principles and innovations shaping organic reaction mechanisms today. Each contribution reflects the expertise of its authors and the ever-evolving demands of contemporary organic chemistry.

Conclusion

As we draw this volume to a close, it is evident that the foundational principles and advanced concepts explored throughout these chapters serve not only as pillars of organic chemistry but also as springboards for continued innovation and application across scientific disciplines. From the critical understanding of chemical reactivity and mechanistic insight into substitution, elimination, and proton-coupled electron transfer, to the practical relevance of reactive intermediates and sophisticated spectroscopic analysis, each contribution reflects the evolving complexity and interdisciplinary nature of modern organic chemistry.

The inclusion of chapters focused on sulfur-based rearrangements and the nuanced understanding of isomerism further underscores the relevance of organic compounds in medicinal chemistry, materials science, and industrial applications. The comprehensive treatment of both classical and emerging topics in this volume equips readers with a deep, integrative understanding essential for navigating contemporary research challenges and for fostering novel discoveries.

As editor, it is my hope that this compilation will serve as both a reference and an inspiration—bridging foundational knowledge with advanced inquiry, and stimulating curiosity and critical thinking among students, educators, and researchers alike. Let this book not be an endpoint but a beginning—a catalyst for deeper exploration, collaboration, and advancement in the ever-dynamic field of organic chemistry.

Chapter 1
Understanding Chemical Reactivity:
Acids, Bases, and Nucleophiles

Rabab M. Nasser
https://orcid.org/0000-0003-2806-1051
Egyptian Petroleum Research Institute, Egypt

ABSTRACT

Understanding the fundamental principles of chemical reactivity is crucial in the field of organic chemistry. This chapter explores three key concepts that govern the reactivity of organic compounds: acids, bases, and nucleophiles. The Brønsted-Lowry theory is introduced, which defines acids as proton donors and bases as proton acceptors. The strength of acids and bases, as measured by their pKa and pKb values, is discussed in detail, highlighting how it determines their ability to participate in proton-transfer reactions. Factors influencing acid and base strength, such as electronegativity, resonances, and inductive effects, were examined. The chapter also covers the role of nucleophiles in organic transformations, emphasizing how their reactivity depends on factors like charge, polarizability, and steric effects. Understanding the interplay between acids, bases, and nucleophiles provides a foundation for predicting and controlling the outcomes of a wide range of organic reactions, enabling the rational design of synthetic strategies and the development of new chemical processes.

DOI: 10.4018/979-8-3693-6473-4.ch001

INTRODUCTION

Importance of Understanding Reactivity In Organic Chemistry

The ability to predict and control the reactivity of organic compounds is fundamental to the field of organic chemistry. Understanding the factors that influence the reactivity of molecules is crucial for the design and optimization of organic reactions, the synthesis of complex molecules, and the development of new chemical processes. Reactivity is at the core of organic transformations, governing the rates, selectivities, and outcomes of chemical reactions. By comprehending the underlying principles of chemical reactivity, organic chemists can devise more efficient synthetic strategies, develop new catalytic systems, and explore novel reactivity patterns, ultimately advancing the field of organic chemistry and its applications in various industries, such as pharmaceuticals, materials science, and sustainable energy.

Overview of Key Concepts: Acids, Bases, And Nucleophiles

Three central concepts that are essential for understanding chemical reactivity in organic chemistry are acids, bases, and nucleophiles. Acids and bases, as defined by the Brønsted-Lowry theory, play a crucial role in proton-transfer reactions, which are ubiquitous in organic transformations. The strength of an acid or base, as measured by its pKa value, determines its ability to donate or accept protons, and this, in turn, influences the reactivity and selectivity of many organic reactions. Nucleophiles, on the other hand, are species that possess unshared pairs of electrons and are capable of donating these electrons to electrophilic centers, driving a wide range of addition and substitution reactions. The reactivity of nucleophiles is dependent on factors such as charge, polarizability, and steric effects. Understanding the interplay between acids, bases, and nucleophiles is essential for predicting and controlling the outcome of organic reactions, which is the foundation for the rational design of synthetic strategies, (Carey & Sundberg, 2007) (Clayden, Greeves, & Warren, 2012) (McMurry, 2016).

ACIDS AND BASES

Brønsted-Lowry Theory Of Acids And Bases

The fundamental understanding of acids and bases in organic chemistry is based on the Brønsted-Lowry theory, which defines acids as proton donors and bases as proton acceptors. According to this theory, an acid-base reaction involves the transfer of a proton (H+) from the acid to the base, forming a conjugate acid-base pair.

Proton Donors And Acceptors

An acid is a substance that can donate a proton (H+) to another species, while a base is a substance that can accept a proton. In the reaction between an acid and a base, the acid loses a proton, and the base gains a proton. For example, in the reaction between acetic acid (CH_3COOH) and water (H_2O), the acetic acid acts as the proton donor, and the water acts as the proton acceptor:

$$CH_3COOH + H_2O \rightleftharpoons CH_3COO^- + H_3O^+$$

The acetate ion (CH_3COO^-) and the hydronium ion (H_3O+) formed in this reaction are known as the conjugate base of the acid and the conjugate acid of the base, respectively.

Conjugate Acid-Base Pairs

When an acid donates a proton to a base, the acid is converted to its conjugate base, and the base is converted to its conjugate acid. These pairs of substances are called conjugate acid-base pairs. The strength of an acid or base is determined by its ability to donate or accept protons, which is quantified by its pKa value. The smaller the pKa value, the stronger the acid, and the larger the pKa value, the stronger the base. For example, consider the reaction between acetic acid (CH3COOH) and ammonia (NH3):

$CH3COOH + NH3 \rightleftharpoons CH3COO^{-} + NH4^{+}$

In this reaction, acetic acid (CH3COOH) is the acid, and ammonia (NH_3) is the base. The conjugate base of the acid is the acetate ion ($CH3COO^{-}$), and the conjugate acid of the base is the ammonium ion (NH_4^{+}). The strength of the acid and base is determined by their respective pKa values. Understanding the concept of conjugate acid-base pairs is crucial for predicting the direction of acid-base reactions, determining the pH of solutions, and understanding the behavior of buffers, which are essential in various applications, such as biochemistry, analytical chemistry, and industrial processes. The Brønsted-Lowry theory provides a deeper understanding of the fundamental nature of acids and bases, allowing chemists to rationalize and predict the outcomes of a wide range of chemical reactions involving proton transfer. **Table 1** indicates the common conjugate acid-base pairs in organic chemistry.

Table 1. Common conjugate acid-base pairs in organic chemistry

Acid	Conjugate Base	pKa
Hydrochloric acid (HCl)	Chloride ion (Cl-)	-7.0
Sulfuric acid (H2SO4)	Bisulfate ion (HSO4-)	-3.0
Nitric acid (HNO3)	Nitrate ion (NO3-)	-1.4
Acetic acid (CH3COOH)	Acetate ion (CH3COO-)	4.8
Benzoic acid (C6H5COOH)	Benzoate ion (C6H5COO-)	4.2
Formic acid (HCOOH)	Formate ion (HCOO-)	3.8
Carbonic acid (H2CO3)	Bicarbonate ion (HCO3-)	6.4
Ammonia (NH3)	Ammonium ion (NH4+)	9.3
Methylamine (CH3NH2)	Methylammonium ion (CH3NH3+)	10.6
Pyridine (C5H5N)	Pyridinium ion (C5H5NH+)	5.2
Phenol (C6H5OH)	Phenoxide ion (C6H5O-)	9.9

Strength of Acids And Bases

The strength of acids and bases is a fundamental concept in organic chemistry that determines their behavior and reactivity. Acids and bases are classified based on their ability to donate or accept protons (H^+), respectively.

pH Scale and pKa Values

The pH scale and pKa values are closely related and fundamental concepts in understanding the behavior of acids and bases in organic chemistry. The pH scale is a measure of the acidity or basicity of a solution, ranging from 0 to 14. pH is defined as the negative logarithm of the hydrogen ion (H+) concentration in a solution, equation (1):

pH=-log[H+] eq.1

A pH value of 7 is considered neutral, where the concentration of H+ and OH- ions are equal. Solutions with pH values less than 7 are considered acidic, while solutions with pH values greater than 7 are considered basic or alkaline. **Table 2** indicates the common acids and bases with its pKa value.

Table 2. Common acids and bases with pKa values

Compound	Formula	pKa value
Acetic acid	CH3COOH	4.75
Benzoic acid	C6H5COOH	4.20
Boric acid	H3BO3	9.15
Carbonic acid	H2CO3	6.35, 10.33
Formic acid	HCOOH	3.75
Hydrocyanic acid	HCN	9.3
Hydrofluoric acid	HF	3.20
Sulphuric acid	H2SO4	-3, 1.99

continued on following page

Table 2. Continued

Compound	Formula	pKa value
Hydrogen sulphide	H2S	7.05, 19
Nitrous acid	HNO2	3.25
Oxalic acid	H2C2O4	1.2, 4.2
Water	H2O	14
Alkyne	CH CH	25
Amine	NH3	35
Alkane	CH4	50
Phenol	C6H5OH	10
Phosphoric acid	H3PO4	2.16, 7.21, 12.32
Sulfurous acid	H2SO3	1.85, 7.2

The strength of an acid or base is quantified by its pKa or pKb value, which represents the negative logarithm of the acid or base dissociation constant. Strong acids have a low pKa value, indicating a greater tendency to donate protons, while weak acids have a higher pKa value and a lower tendency to donate protons, equation (2).

$pKa = -\log[Ka]$ eq. 2

Where:

pKa is the negative logarithm of the acid dissociation constant (Ka)

Ka is the acid dissociation constant, which represents the extent to which an acid dissociates in water to produce H+ ions.

The acid dissociation constant (Ka) is defined as in equation (3):

$$Ka = \frac{[H+][A-]}{[HA]}$$

eq. (3)

Where:

[H+] is the concentration of hydrogen ions (protons)

[A-] is the concentration of the conjugate base

[HA] is the concentration of the undissociated acid

pKb can be calculated from equation (4):

$$pKb = -\log[Kb]$$

eq. (4)

Where:

pKb is the negative logarithm of the base dissociation constant (Kb)

Kb is the base dissociation constant, which represents the extent to which a base dissociates in water to produce OH^- ions.

The base dissociation constant (Kb) is defined as in equation (5):

$$Kb = \frac{[OH-][HB+]}{[B]} \text{ eq. 5}$$

Where:

[OH-] is the concentration of hydroxide ions

[HB+] is the concentration of the conjugate acid

[B] is the concentration of the undissociated base

Similarly, strong bases have a high pKb value, meaning they are more readily able to accept protons, while weak bases have a lower pKb value and a reduced ability to accept protons.

The strength of an acid or base is inversely related to the strength of its conjugate base or acid, respectively. This relationship, known as the principle of conjugate acid-base pairs, is essential for understanding and predicting the outcomes of acid-base reactions.

Factors Affecting Acid/Base Strength

Electronegativity

Electronegativity is a fundamental property that plays a crucial role in understanding the behavior of atoms and molecules. This property reflects the ability of an atom to attract shared electrons in a chemical bond. Understanding the trends in electronegativity across the periodic table is essential for predicting and explaining chemical reactivity.

When examining the periodic table, we observe distinct trends in electronegativity. As one moves across a period, from left to right, the electronegativity of the elements generally increases. This can be attributed to the increasing number of protons in the nucleus, which results in a stronger pull on the valence electrons. For example, the electronegativity values across the second period are: lithium (0.98), beryllium (1.57), boron (2.04), carbon (2.55), nitrogen (3.04), oxygen (3.44), and fluorine (3.98). The steady increase in electronegativity from left to right in the periodic table is a consequence of the growing nuclear charge and the decreasing atomic radius.

In contrast, when moving down a group, or family, of the periodic table, the electronegativity of the elements typically decreases. This trend can be explained by the increasing atomic radius as you descend a group. As the valence electrons become farther from the nucleus, the pull on these electrons weakens, resulting in a lower electronegativity. The halogens, for instance, exhibit this trend, with fluorine having the highest electronegativity (3.98), followed by chlorine (3.16), bromine (2.96), and iodine (2.66).

To further illustrate the trends in electronegativity, we can arrange the elements and groups in **Table 3** according to their electronegativity values. This table highlights the gradual changes in electronegativity across the periodic table.

Electronegativity is directly related to the pKa of an acid. The higher the electronegativity of the atom attached to the hydrogen in an acid, the lower the pKa (and the stronger the acid). This is because the high electronegativity of the atom pulls electron density away from the hydrogen, making it more likely to dissociate as a proton (H+).

Table 3. Arrangement of elements and groups based on electronegativity values

Electronegativity Range	Elements/Groups
4.0 - 4.0	Fluorine (F)
3.5 - 3.9	Oxygen (O), Chlorine (Cl)
3.0 - 3.4	Nitrogen (N), Bromine (Br)
2.5 - 2.9	Sulfur (S), Iodine (I)
2.0 - 2.4	Phosphorus (P), Silicon (Si), Arsenic (As)
1.5 - 1.9	Aluminum (Al), Germanium (Ge), Tin (Sn), Lead (Pb)
1.0 - 1.4	Magnesium (Mg), Calcium (Ca), Strontium (Sr), Barium (Ba)

For Example, Acetic acid (CH_3COOH) has a pKa of 4.76, while trifluoroacetic acid (CF_3COOH) has a pKa of 0.23. The higher electronegativity of the fluorine atoms in trifluoroacetic acid compared to the hydrogen atoms in acetic acid makes the former a stronger acid.

Resonance Stabilization

One of the important factors that can significantly influence the strength of an acid or base is the concept of resonance stabilization. Resonance stabilization refers to the ability of a molecule or ion to spread the distribution of electrons across multiple equivalent structures, known as resonance structures, thereby reducing the overall energy of the system.

When a molecule or ion exhibits resonance stabilization, the positive or negative charge associated with the acid or base is delocalized across multiple atoms, rather than being concentrated on a single atom. This delocalization of charge results in a more stable configuration, which in turn affects the acid or base strength.

For acids, resonance stabilization of the conjugate base typically enhances the acid's strength. The more the conjugate base is stabilized through resonance, the less it will seek to regain the proton, and the stronger the original acid will be. Conversely, if the conjugate base is not resonance-stabilized, the acid will be weaker, as the conjugate base will be more inclined to recapture the proton.

For bases, resonance stabilization of the conjugate acid can enhance the base's strength. If the conjugate acid is resonance-stabilized, it will be less inclined to release the proton, and the original base will be a stronger base.

Some examples of how resonance stabilization affects acid/base strength include:

- **Carboxylic acids (R-COOH):** The conjugate base of a carboxylic acid, the carboxylate ion (R-COO$^-$), is resonance-stabilized, making the original carboxylic acid a stronger acid.
- **Phenols (ArOH):** The conjugate base of a phenol, the phenoxide ion (ArO$^-$), is resonance-stabilized, contributing to the relatively high acidity of phenols.
- **Pyridines (C5H5N):** The conjugate acid of pyridine, the pyridinium ion ($C_5H_5NH^+$), is resonance-stabilized, making pyridine a relatively weak base.

Inductive Effects

In addition to resonance stabilization, the inductive effect is another important factor that can influence the strength of acids and bases. The inductive effect refers to the ability of atoms or functional groups to either withdraw or donate electron density through the sigma (σ) bonds of a molecule.

The inductive effect can have a significant impact on the acidity or basicity of a compound, as it affects the stability of the conjugate acid or base. Electron-withdrawing groups (EWGs) have a tendency to pull electron density away from the atom or functional group they are attached to. This stabilizes the conjugate base, making the original compound a stronger acid. Conversely, electron-donating groups (EDGs) push electron density towards the atom or functional group they are attached to, destabilizing the conjugate base and making the original compound a weaker acid. For example, consider the acidity of acetic acid (CH_3COOH) and trifluoroacetic acid (CF_3COOH). The presence of the three electronegative fluorine atoms in trifluoroacetic acid creates a strong electron-withdrawing inductive effect, which stabilizes the conjugate base (CF_3COO^-) and makes trifluoroacetic acid a stronger acid compared to acetic acid (CH_3COOH).

Similarly, the inductive effect can also influence the basicity of a compound. Electron-withdrawing groups (EWGs) tend to destabilize the conjugate acid, making the original compound a weaker base. Conversely, electron-donating groups (EDGs) can stabilize the conjugate acid, resulting in a stronger base. For instance, the basicity of pyridine (C_5H_5N) is influenced by the presence of the electronegative nitrogen atom, which has an electron-withdrawing inductive effect. This destabilizes the conjugate acid (the pyridinium ion, $C_5H_5NH^+$), making pyridine a relatively weak base compared to other amines.

It's important to note that the inductive effect is a localized phenomenon, and its impact on acidity or basicity can be influenced by the distance of the substituent from the site of protonation or deprotonation. Generally, the closer the electron-withdrawing or electron-donating group is to the acid or base site, the more pronounced its effect will be.

Acid-Base Equilibria And Reactions

Acid-Base Reactions

Acid-base reactions involve the transfer of protons (H^+) between chemical species. These reactions can be described using the Brønsted-Lowry definition, where an acid is a proton donor, and a base is a proton acceptor. The general equation for an acid-base reaction can be written as:

$$\text{Acid} + \text{Base} \rightleftharpoons \text{Conjugate Base} + \text{Conjugate Acid}$$

The strength of an acid or base is determined by its propensity to donate or accept protons, respectively. The acid dissociation constant (Ka) and the base dissociation constant (Kb) are used to quantify the strength of acids and bases, respectively. These constants are related through the water autoionization constant ($Kw = 1.0 \times 10^{-14}$ at 25°C), (Atkins & de Paula, 2014).

Calculating pH of Solutions

The pH of a solution is a measure of the concentration of hydrogen ions (H^+) in the solution, and it is defined as the negative logarithm of the hydrogen ion concentration: pH = -log [H^+]. The pH of a solution can be calculated using the following approaches:

- Strong Acids and Bases:

For strong acids and bases, the pH can be determined directly from the concentration of the acid or base.

- Weak Acids and Bases:

For weak acids and bases, the pH is determined by the equilibrium between the acid/base and its conjugate species. The pH can be calculated using the acid or base dissociation constant (Ka or Kb) and the initial concentration of the acid or base.

- Polyprotic Acids:

Polyprotic acids, such as sulfuric acid (H_2SO_4) and phosphoric acid (H_3PO_4), have multiple dissociation steps and require more complex calculations to determine the pH, (Chang & Goldsby, 2013).

Buffers and pH Control

Buffers are essential for maintaining a specific pH in various applications, such as in biological systems, chemical reactions, and industrial processes. The pH of a buffer solution can be adjusted by changing the ratio of the conjugate acid-base pair or by selecting a buffer system with a different pKa value, (Silberbeg & Amateis, 2018).

NUCLEOPHILES AND ELECTROPHILES

Definition and Characteristics of Nucleophiles

Nucleophiles are species that are attracted to areas of low electron density and have the ability to donate electron pairs to form new chemical bonds. They are essential participants in many organic reactions, such as additions, substitutions, and eliminations. Nucleophiles can be neutral molecules, anions, or even cations in certain cases. Some common examples of nucleophiles include water (H_2O), hydroxide ion (OH^-), ammonia (NH_3), and alkoxide ions (RO^-), **Table 4.**

Table 4. Common nucleophiles and their relative reactivities

Nucleophile	Relative Reactivity
Hydroxide Ion (OH-)	High
Water (H2O)	Moderate
Amine (NH3)	High
Thiolate Ions	Moderate
Alkoxide Ions	High
Halide Ions (F-, Cl-, Br-)	Moderate to High
Cyanide Ion (CN-)	High
Enolate Ions	High

The defining characteristic of a nucleophile is its ability to donate a pair of electrons to form a new covalent bond. Nucleophiles can be classified based on their atom type, which includes carbon, oxygen, nitrogen, sulfur, and halogen nucleophiles. The strength of a nucleophile is determined by its ability to donate electrons, which is influenced by factors such as charge, polarizability, and steric effects.

Factors Affecting Nucleophilicity

- *Charge*

The charge on a nucleophile plays a significant role in its reactivity. Generally, nucleophiles with a negative charge (anionic nucleophiles) are more reactive than neutral nucleophiles. This is because the negative charge increases the electron density and makes the nucleophile more electron-rich, enhancing its ability to donate electrons. For example, the hydroxide ion (OH^-) is a stronger nucleophile than water (H_2O) due to its negative charge.

- *Polarizability*

Polarizability refers to the ease with which the electron cloud of a nucleophile can be distorted or polarized. Nucleophiles with higher polarizability are generally more reactive. Larger atoms, such as iodide (I^-) and thiolate (RS^-), have more easily polarizable electron clouds compared to smaller atoms, such as fluoride (F^-) and alkoxide (RO^-) ions. The increased polarizability enhances the nucleophile's ability to form new bonds.

- *Steric Effects*

Steric effects refer to the influence of the size and shape of a nucleophile on its reactivity. Bulky or sterically hindered nucleophiles have reduced reactivity due to the increased difficulty in approaching the electrophilic site. For example, the tertiary butoxide ion ($t\text{-}BuO^-$) is a less reactive nucleophile compared to the smaller methoxide ion (CH_3O^-) due to the bulky tert-butyl group.

Electrophiles and Their Reactivity

Electrophiles are species that are attracted to areas of high electron density and have the ability to accept electron pairs to form new chemical bonds. They are commonly involved in electrophilic addition and substitution reactions. Electrophiles can be neutral molecules, cations, or even certain neutral atoms, such as boron (B) and silicon (Si). Examples of electrophiles include protons (H^+), carbocation's (R_3C^+), and certain metal ions (e.g., Fe^{3+}, Cu^{2+}).

Electrophilic Additions and Substitutions

Electrophilic additions involve the addition of an electrophile to an unsaturated compound, such as an alkene or an aromatic ring. The electrophile attacks the electron-rich double bond or aromatic system, leading to the formation of a new product. Electrophilic substitutions, on the other hand, involve the replacement of a substituent in a molecule by an electrophilic species.

Factors Affecting Electrophilicity

The reactivity of an electrophile is primarily influenced by the following factors:

- *Charge:* Positively charged electrophiles are generally more reactive than neutral electrophiles due to the increased electron deficiency and ability to accept electron pairs.
- *Polarizability:* Larger and more polarizable electrophiles are typically more reactive. This is because the electron cloud of the electrophile can be more easily distorted, facilitating the formation of a new bond.
- Solvation: The degree of solvation of an electrophile can affect its reactivity. Electrophiles that are well-solvated tend to be less reactive, as the solvation shell must be disrupted for the electrophile to interact with the nucleophile.
- Stability of the Transition State: Electrophiles that can stabilize the transition state of the reaction are more reactive. This is often achieved through resonance stabilization or the formation of more stable carbocation.

Understanding the factors that influence nucleophilicity and electrophilicity is crucial for predicting and explaining the reactivity of chemical species in organic reactions. By considering these principles, chemists can design effective synthetic strategies and better understand the underlying mechanisms of various organic transformations.

APPLICATIONS IN ORGANIC REACTIONS

Nucleophilic Substitution Reactions

Nucleophilic substitution reactions are fundamental organic transformations in which a nucleophile replaces a leaving group, such as a halide or a sulfonate ester, in a substrate molecule. There are two main mechanisms for these reactions: the SN1 (Substitution Nucleophilic Unimolecular) and the SN2 (Substitution Nucleophilic Bimolecular) mechanisms.

Figure 1. General mechanism of electrophilic addition reaction

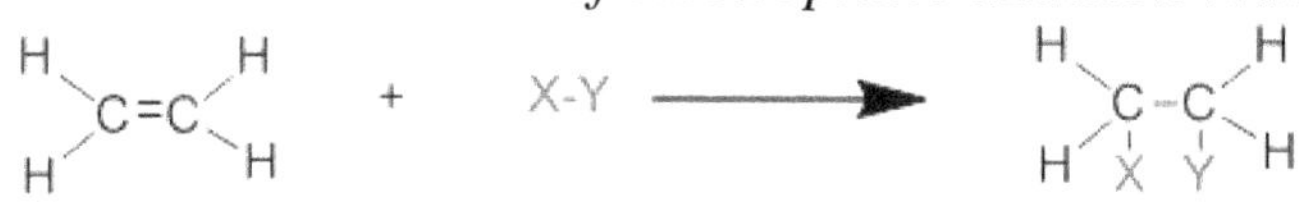

SN^1 and SN^2 Mechanisms

Figure 2. Mechanism of SN1 reaction

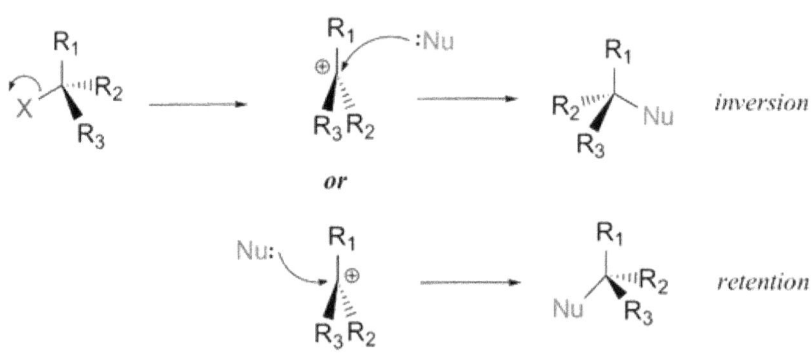

Figure 3. Energy diagram of SN1 reaction

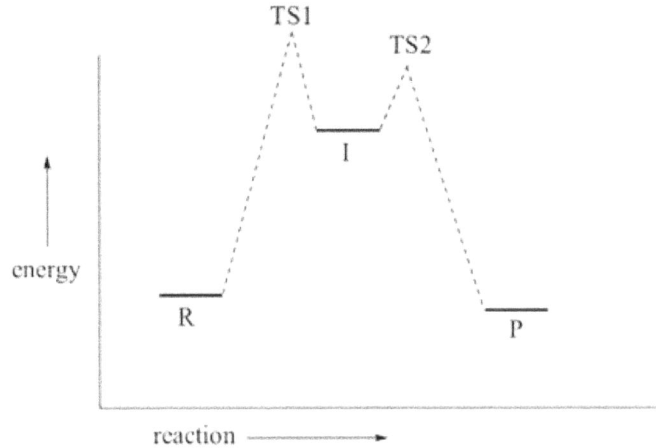

The SN1 mechanism involves a two-step process, where the first step is the formation of a carbocation intermediate, followed by the attack of the nucleophile. This mechanism is favored in substrates with tertiary or benzylic carbocation, which are more stable, **Figures 2 and 3**. The SN2 mechanism, on the other hand, is a one-step process where the nucleophile attacks the carbon bearing the leaving group in a backside attack, leading to inversion of stereochemistry, **Figures 4 and 5**.

Figure 4. SN2 mechanism

Figure 5. Energy diagram of SN2 reaction

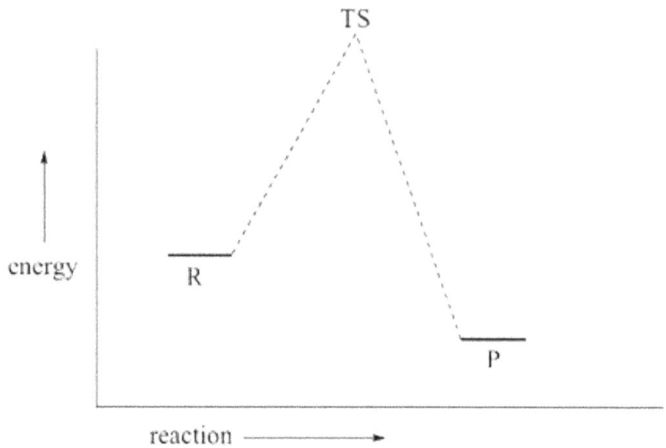

Factors Affecting the SN1 and SN2 Pathways

The predominant reaction pathway (SN1 or SN2) is influenced by several factors, including the nature of the nucleophile, the substrate, and the leaving group.

- *Nature of the Nucleophile:*

Nucleophiles with high nucleophilicity (e.g., alkoxide ions, thiolates) generally favor the SN2 pathway. Weaker nucleophiles (e.g., water, alcohols) tend to follow the SN1 pathway.

- *Nature of the Substrate:*

Tertiary and benzylic substrates are more likely to undergo SN1 reactions due to the stability of the carbocation intermediate. Primary and secondary substrates typically favor the SN2 mechanism.

- *Nature of the Leaving Group:*

Better leaving groups (e.g., halides, sulfonate esters) promote the SN1 pathway by facilitating the formation of the carbocation intermediate. Poorer leaving groups (e.g., alcohols, amines) generally lead to the SN2 mechanism.

Nucleophilic Addition Reactions

Nucleophilic addition reactions involve the addition of a nucleophile to an electrophilic carbon-containing functional group, such as a carbonyl compound (aldehyde, ketone, ester, etc.).

Addition to Carbonyl Compounds

Nucleophilic addition to carbonyl compounds is a common reaction in organic chemistry. The carbonyl carbon is electrophilic due to the partial positive charge, and the nucleophile attacks the carbonyl carbon to form a tetrahedral intermediate, which then collapses to form the final product, **Figure 6**.

Figure 6. Addition to carbonyl compounds

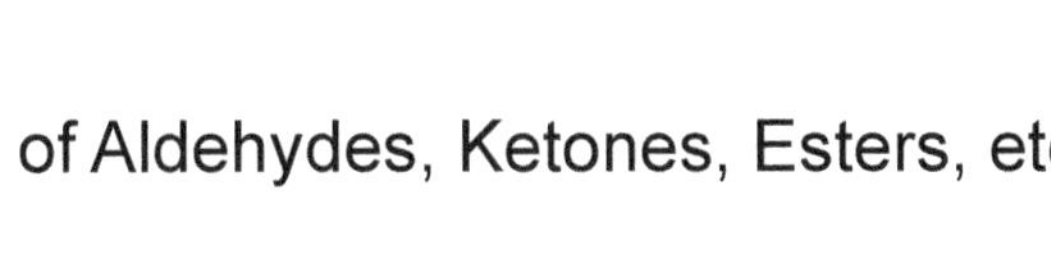

Reactivity of Aldehydes, Ketones, Esters, etc.

The reactivity of carbonyl compounds in nucleophilic addition reactions can be influenced by factors such as the hybridization of the carbonyl carbon, the presence of electron-withdrawing or electron-donating substituents, and the overall stability of the intermediate species formed. For example, aldehydes are generally more reactive than ketones in nucleophilic addition reactions due to the reduced steric hindrance and the greater electrophilicity of the carbonyl carbon. Esters, on the other hand, are less reactive compared to aldehydes and ketones due to the stabilizing effect of the alkoxy group.

Acid-Base Catalysis in Organic Reactions

Acid-base catalysis plays a crucial role in many organic reactions, particularly in proton transfers and the formation and reactions of enolates.

Proton Transfers

Proton transfers are essential steps in various organic reactions, such as esterification, hydrolysis, and tautomerization. These processes are often facilitated by the presence of acids or bases, which can either donate or accept protons, thereby lowering the activation energy and increasing the rate of the reaction.

Enolate Formation and Reactions

Figure 7. Keto-enol form

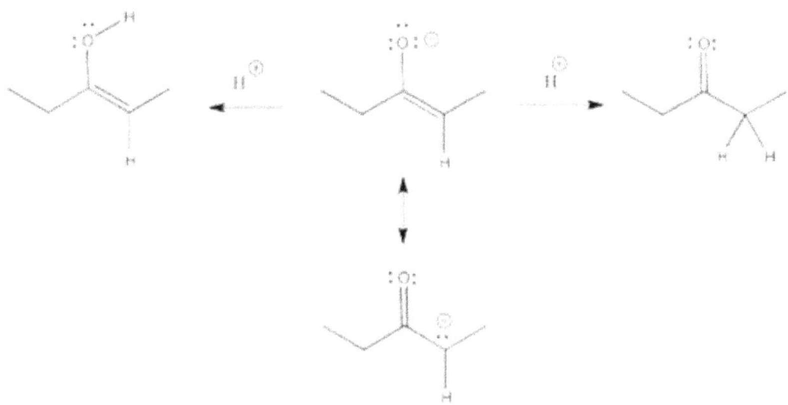

Enolate formation is an important reaction in organic chemistry, particularly in the context of carbonyl compound reactivity. An enolate is the conjugate base of an enol, which is the tautomeric form of a carbonyl compound where the carbonyl carbon is connected to a hydroxyl group, **Figure 7**. Enolate formation occurs when a carbonyl compound, such as a ketone or aldehyde, is treated with a strong base, typically an alkoxide or amide base. The base removes a proton from the α-carbon (the carbon atom adjacent to the carbonyl), resulting in the formation of a planar, negatively charged enolate ion. Enolates are highly reactive nucleophiles that can participate in a variety of organic reactions, including aldol condensations, Claisen condensations, and enolate alkylations. The ability to generate and control enolate intermediates is a fundamental skill in organic synthesis, as it allows for the introduction of new carbon-carbon bonds and the construction of more complex molecular structures.

EXERCISES

I. Identify the acid, base, conjugate acid, and conjugate base in the reaction: NH3 + HCl ⇌ NH4+ + Cl-
II. Determine the conjugate acid-base pairs in the reaction: CH3COOH + H2O ⇌ CH3COO- + H3O+
III. Calculate the pH of a solution containing 0.1 M acetic acid (CH3COOH, pKa = 4.76) and 0.2 M sodium acetate (CH3COONa).
IV. Rank the following acids in order of increasing strength: HF, HCl, HBr, HI.
V. Explain how the electronegativity of the central atom in an acid affects its strength.
VI. Predict the relative strengths of the conjugate bases of H2SO4 and H3PO4.
VII. Determine the most nucleophilic species among the following: H2O, NH3, CH3O-, Cl-.
VIII. Explain how the charge and polarizability of a nucleophile affect its reactivity.
IX. Propose a mechanism for the reaction between ethyl bromide (CH3CH2Br) and hydroxide ion (OH-) to form ethanol (CH3CH2OH).

X. A solution is prepared by mixing equal volumes of 0.1 M acetic acid (CH3COOH) and 0.1 M ammonia (NH3). Calculate the pH of the resulting solution.
XI. Predict the products of the reaction between benzoic acid (C6H5COOH) and sodium hydroxide (NaOH). Write the balanced equation and identify the conjugate acid-base pairs.
XII. Propose a mechanism for the reaction between 2-bromobutane and sodium ethoxide (CH3CH2O-) to form 2-ethoxybutane. Discuss the factors that influence the reactivity of the nucleophile and the leaving group.

CONCLUSION

The ability to understand and predict the reactivity of organic compounds is fundamental to the field of organic chemistry. This chapter has explored three central concepts - acids, bases, and nucleophiles - that are essential for comprehending the underlying principles of chemical reactivity. By mastering the Brønsted-Lowry theory of acid-base interactions, the factors that influence the strength of acids and bases, and the characteristics that govern nucleophilic reactivity, organic chemists can devise more efficient synthetic pathways, develop new catalytic systems, and explore novel reactivity patterns. This knowledge forms the foundation for the rational design of organic reactions, the synthesis of complex molecules, and the advancement of various applications in areas such as pharmaceuticals, materials science, and sustainable energy. Continued exploration and understanding of chemical reactivity will undoubtedly drive further progress in the field of organic chemistry and its diverse real-world applications.

REFERENCES

Atkins, P., & de Paula, J. (2014). *Atkins' Physical Chemistry* (10th ed.). Oxford University Press. DOI: 10.1093/acprof:oso/9780198769286.001.0001

Carey, F. A., & Sundberg, R. J. (2007). *Advanced Organic Chemistry: Part A: Structure and Mechanisms*. Springer Science & Business Media. DOI: 10.1007/978-0-387-44899-2

Chang, R., & Goldsby, K. A. (2013). *Chemistry* (11th ed.). McGraw-Hill Education. DOI: 10.1036/0073402661

Clayden, J., Greeves, N., & Warren, S. (2012). *Organic Chemistry*. Oxford University Press. DOI: 10.1093/hesc/9780199270293.001.0001

McMurry, J. E. (2016). *Organic Chemistry*. Cengage Learning. DOI: 10.1016/B978-0-12-407721-2.00018-7

Silberberg, M. S., & Amateis, P. (2018). *Chemistry: The Molecular Nature of Matter and Change* (8th ed.). McGraw-Hill Education. DOI: 10.1036/0078747872

Chapter 2
Organohalides Substitution and Elimination Reactions

Khadra B. Alomari
https://orcid.org/0000-0001-5995-0172
Jazan University, Saudi Arabia

ABSTRACT

Substitution and elimination are considered the most important types of reactions in organic chemistry. Substitution reactions are chemical reactions that involve the replacement of one functional group in a molecule or ion with another functional group. During the substitution, the bond between the functional group and the reactive centre is broken, and a new bond forms between that centre and the new functional group. On the other hand, elimination reactions are a type of organic chemical reaction in which a pair of atoms or groups are removed from a molecule. It is the primary process by which organic compounds with only one carbon-carbon bond (saturated compounds) are converted to compounds with two or three carbon-carbon bonds (unsaturated compounds). It is the primary process by which organic compounds with only one carbon-carbon bond (saturated compounds) are converted to compounds with two or three carbon-carbon bonds (unsaturated compounds).

DOI: 10.4018/979-8-3693-6473-4.ch002

INTRODUCTION TO SUBSTITUTION AND ELIMINATION REACTIONS

This chapter will specifically examine the reactions of alkyl halides, which are compounds that contain halogens (such as Cl, Br, I) bonded to a carbon atom that is sp^3 hybridized. Alkyl halides typically undergo two broad categories of reactions (Figure 1) (Murry, 1977). When subjected to a nucleophile, an alkyl halide has the ability to undergo a substitution reaction, wherein the halogen is replaced by the nucleophile. When subjected to a base, an alkyl halide has the potential to undergo an elimination reaction, resulting in the formation of a *pi* bond, specifically an alkene.

Figure 1. Possible reactions of alkyl halide

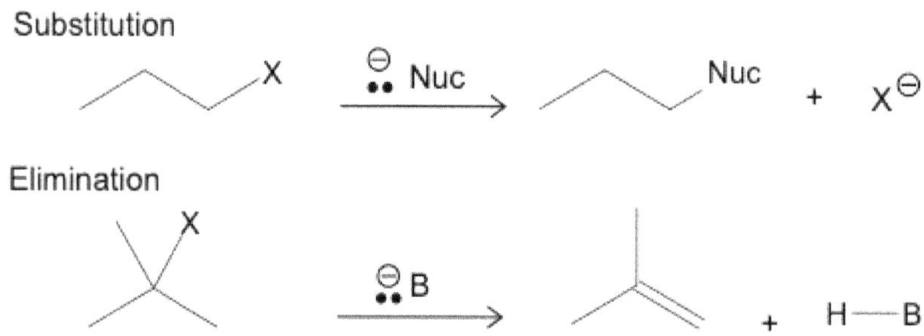

Due to the dual nature of certain reagents, such as hydroxide ion (OH^-), they can act as both nucleophiles and bases. As a result, substitution and elimination reactions frequently occur simultaneously, leading to competition between the two, as demonstrated in the example (Figure 2) (Steigman, 1937). This chapter will examine substitution and elimination reactions, as well as the factors that determine the competition between these reactions.

Figure 2. Competition between substitution and elimination reactions

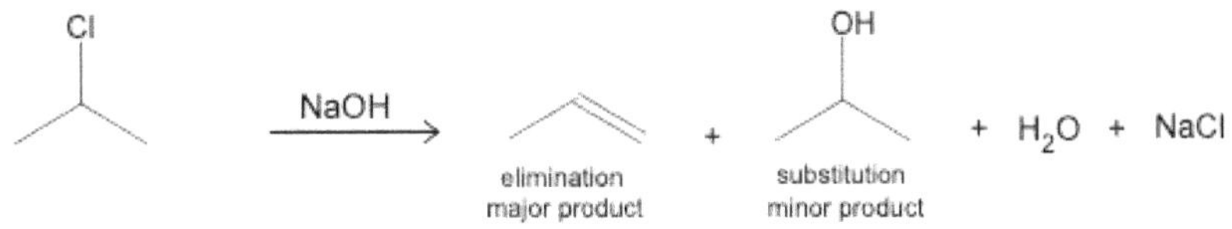

The halogen in alkyl halide plays two important roles that make the alkyl halide susceptible to substitution and elimination reactions. The halogen atom removes electron density through induction, making the neighbouring carbon atom electron-deficient and thus susceptible to nucleophilic attack. The halogen can function as a leaving group, and the process of substitution and elimination can only take place in the presence of a leaving group. A good leaving group is the conjugate base of a strong acid. Among the long halogens, iodide is considered the most favourable leaving group due to its low basicity.

Structure of Alkyl Halides

Each carbon atom in an alkyl halide can be characterized based on its proximity to the halogen using Greek alphabet letters. The alpha (α) position refers to the carbon atoms that are directly bonded to the halogen, whereas the beta (β) positions are the carbon atoms that are attached to the alpha (α) position (Figure 3). An alkyl halide possesses a solitary alpha (α) position, while it can have up to three beta (β) positions. Alkyl halides are categorized as primary (1°), secondary (2°), or tertiary (3°) depending on the quantity of alkyl groups attached to the alpha (α) position.

Figure 3. Classification of alkyl halides scheme

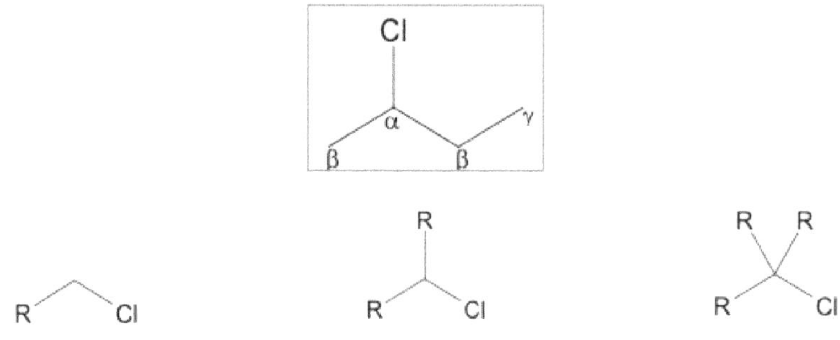

Primary (1°) **Secondary (2°)** **Tertiary (3°)**

Naming Alkyl Halides

While they are often referred to as alkyl halides, halogen-substituted alkenes are more accurately named haloalkanes according to the systematic naming convention (IUPAC name). In this naming system, the halogen is considered a substituent on a parent alkane chain. (Kevill, 1959) There are a total of three sequential steps.

First step: Determine the chain with the greatest length and designate it as the parent. The parent chain must include the multiple bond if it is present.

Second step: Assign numbers to the carbon atoms of the parent chain, starting from the end that is closer to the first substituent, which can be an alkyl or halo group. Number each substituent based on its position along the chain. When there are substituents equidistant from both ends, start numbering from the end that is closer to the substituent with higher alphabetical priority (Figure 4).

Figure 4. IUPAC names of haloalkenes

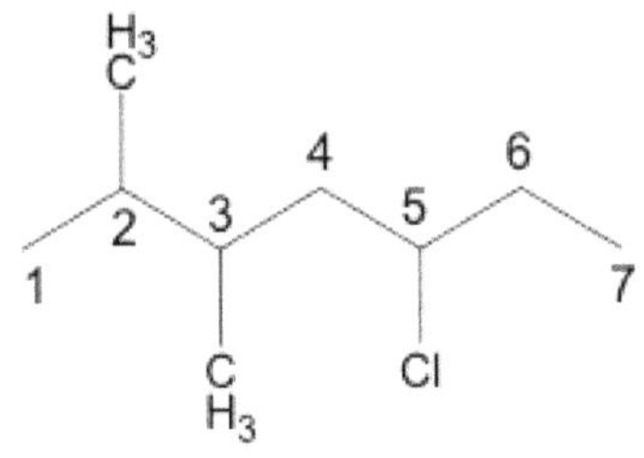

5-Chloro-2,3-dimethylheptane

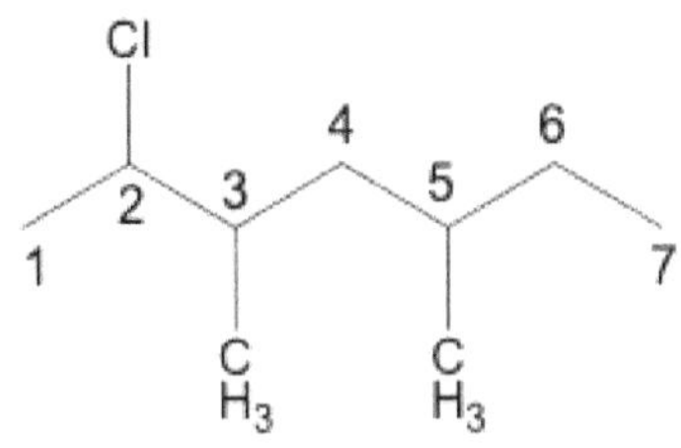

2-Chloro-3,5-dimethylheptane

Third step: Record the name. Enumerate all substituents in alphabetical order, and employ the prefixes di-, tri-, and so on, if multiple instances of the same substituent are present.

Simple alkanes are named not only by their systematic names but also by first identifying the alkyl group and the halogen. For instance, the compound CH_3Br can be referred to as either bromomethane or methyl bromide. Although these names are widely used in the chemical literature and in everyday language, they will not be utilized in this context.

Uses of Organohalides

Organic compounds with halogen substituents are abundant in nature, with over 5000 organohalides discovered in algae and other marine organisms. (Robertson, 2018) Chloromethane, a significant quantity of which is emitted by ocean kelp, forest fires, and volcanoes, undergoes consumption by various bacteria and is subsequently transformed into carbon dioxide and chlorine. Halogen-containing compounds have a wide range of industrial applications, such as being used as solvents, inhaled anesthetics, refrigerants, and pesticides. The contemporary electronics industry heavily depends on halogenated solvents, such as trichloroethylene, for the purpose of cleaning semiconductor chips and other compounds (Figure 5).

Figure 5. Uses of organohalides

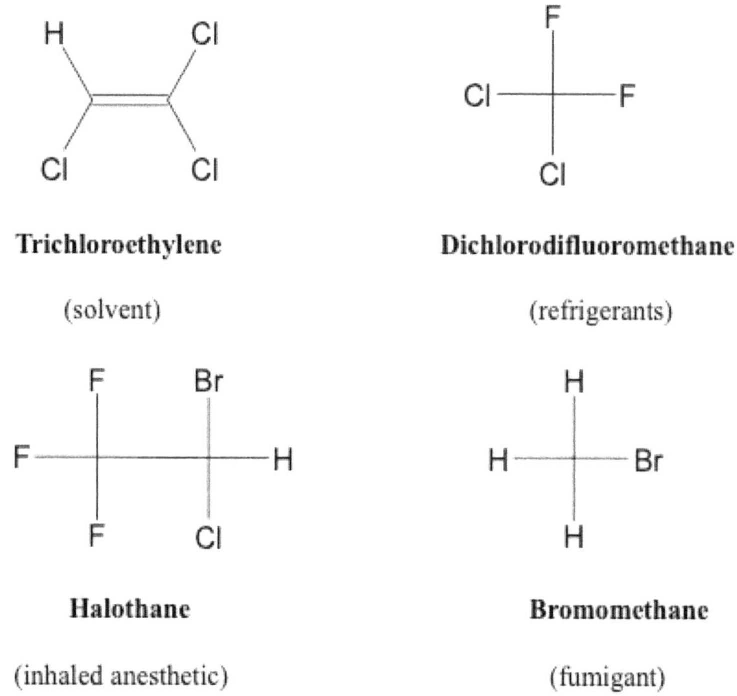

Preparation of Alkyl Halides

The incorporation of halogen acids, represented by formula HX, into alkenes, is a widely applicable chemical reaction that enables chemists to synthesize a diverse range of alkane products substituted with halogens (Figure 4) (Narain, 2011). Therefore, alkenes can undergo additional reactions with HCl, HBr, and HI. Upon careful examination of the product, it was observed that an unsymmetrically substituted alkene yielded a singular addition product, rather than the anticipated mixture. The result is due to the regiospecific nature of the reaction, where only one of the two possible addition products, known as the Markonikov's product (product **1**, Figure 6), is formed. In 1869, the Russian chemist Vladimir Markonikov put forward a principle that has since been referred to as Markonikov's rule. (Robertson, 2018) According to this rule, when HX is added to an alkene, the hydrogen atom attaches to the carbon atom with fewer alkyl substituents, while the halogen atom attaches to the carbon atom with more alkyl substituents. When both carbon atoms with double bonds have an equal level of substitution, it leads to the formation of a mixture of additional products. The reactions occur in a biphasic manner and involve the formation of a carbocation intermediate.

Figure 6. Preparing alkyl halides using addition of HX to alkene method

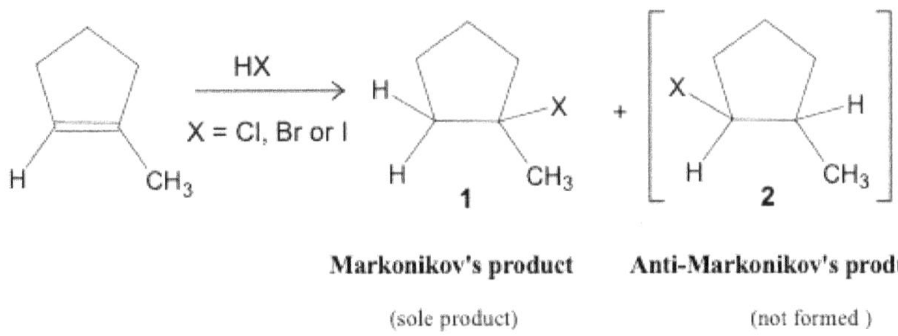

In the process of halogenation, the addition of halogens (X_2) to alkenes can result in the formation of 1,2-dihaloalkenes (Figure 7). When cycloalkenes, like cyclopentene, undergo halogenation, only *trans*-1,2-dihalocyclopentane is produced (**1**). This is surprising because one would expect a mixture of *trans* and *cis* products if a planar carbocation intermediate were involved. Indeed, the reaction proceeds with anti-stereochemistry, indicating that the two halogen atoms originate from opposite sides of the double bond, separated by approximately 180 degrees.

Figure 7. Preparing alkyl halides using addition X2 to alkene method

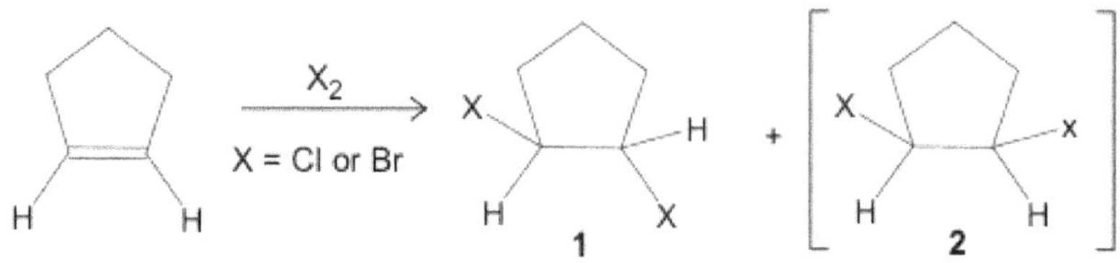

trans-1,2-dihalocyclopentane (sole product)   *cis*-1,2-dihalocyclopentane (not formed)

The stereochemical outcome can be most effectively elucidated by envisioning that the reaction intermediate is not a genuine carbocation, but rather a halonium ion, R_2X^+ (Figure 8), which is formed in a single step through the addition of X^+ to the double bond. Due to the effective shielding of one side of the molecule by the halogen atom, the reaction with the halogen ion X^- in the second step takes place from the opposite side, which is more easily accessible, resulting in the formation of the anti-product.

Figure 8. Mechanism of the addition of X2 to an alkene

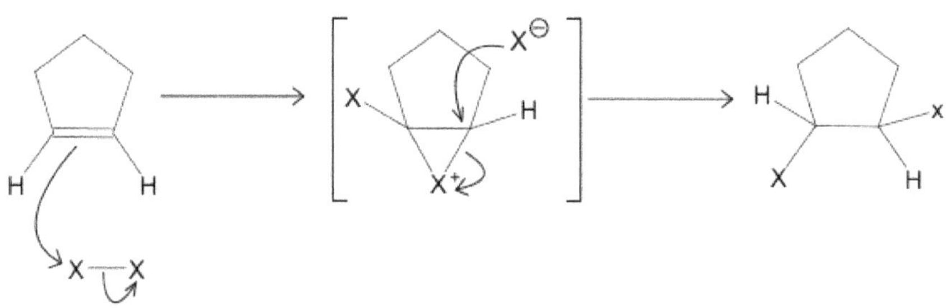

Over 10 million tons of 1,2-dichloroethane, also known as ethylene dichloride, are produced annually in the United States through the process of adding chlorine gas (Cl_2) to ethene. The product serves dual purposes as a solvent and as a primary component for the synthesis of poly (vinyl chloride), PVC. The addition of Br_2 serves as a quick and straightforward laboratory test for unsaturation. A specimen of unknown composition is dissolved in dichloromethane, and a few drops of bromine are subsequently introduced. A rapid vanishing of the reddish bromine hue indicates a positive test and signifies the presence of a double bond in the sample molecule.

The most commonly employed technique for synthesizing alkyl halides involves their conversion from alcohols, which can be readily derived from carbonyl compounds. The reaction can frequently be conducted by directly treating the alcohol with hydrochloric acid (HCl) or hydrobromic acid (HBr). The compound 1-methylcyclohexanol can be transformed into 1-chloro-1-methylcyclohexane through a reaction with hydrochloric acid (HCl) (Figure 9). The HX reaction is most effective when used with tertiary alcohols, as will be explained further in this chapter. Primary and secondary alcohols exhibit a significantly lower reaction rate. Thionyl chloride ($SOCl_2$) or phosphorus tribromide (PBr_3) are the most effective reagents for converting primary and secondary alcohols into alkyl halides (Figure 9). Typically, these reactions occur with high efficiency, resulting in a significant amount of product.

Figure 9. Preparing alkyl halide by replacing alcohol

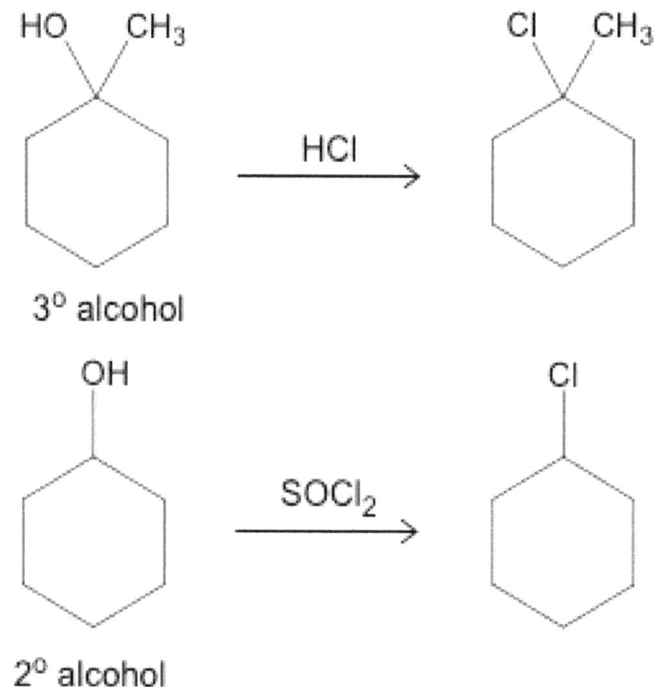

Alkyl fluoride can be derived from alcohol. Various alternative reagents, such as diethylamino sulfur trifluoride and HF-pyridine, are employed for the reaction. Pyridine, being the nitrogen-containing counterpart of benzene, is also utilized.

NUCLEOPHILIC SUBSTITUTION REACTION

In 1896, German scientist Paul Walden made the first discovery of the nucleophilic substitution reaction of alkyl halides. He found that (+)- and (-)-malic acids could be converted into each other. (Solomons et al., 2016) Walden acquired (+)-chlorosuccinic acid through the reaction of (-)-malic acid with PCl_5. When exposed to wet Ag_2O, this substance underwent a reaction and formed (+)-malic acid. Similarly, the combination of (+)-malic acid and PCl_5 resulted in the formation of (-)-chlorosuccinic acid. This compound was then transformed into (-)-malic acid by treating it with moist Ag_2O. Figure 10 displays the complete sequence of reactions documented by Walden.

Figure 10. Walden's cycle reactions

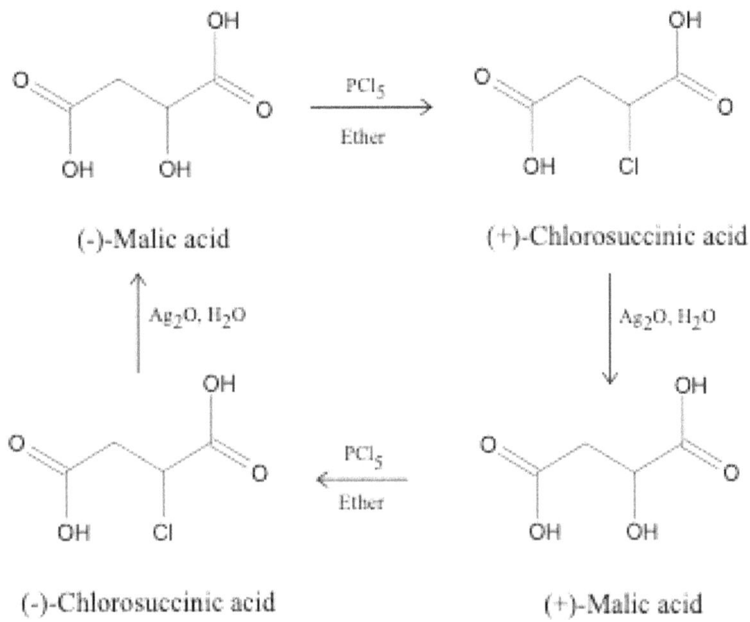

At that moment, the outcomes were astounding. Based on the transformation of (-)-malic acid into (+)-malic acid, it can be deduced that specific reactions in the cycle underwent a reversal or modification in the configuration of the chirality centre. Today, we use the term "nucleophilic substitute reactions" to describe the changes occurring in Walden's cycle. This is because each step of the cycle involves the replacement of one nucleophile (either chlorine, Cl^-, or hydroxide ion, OH^-) with another. Nucleophilic substitution reactions are prevalent and flexible reaction types in the field of organic chemistry.

In a nucleophilic substitution reaction as displayed in Figure 11, a nucleophile (Nu) replaces a leaving group (LG or X^-) in the substrate molecule. A nucleophile is invariably a Lewis base, meaning it donates electron pairs, and it can exist in either a negatively charged or neutral state. The leaving group is invariably a species that carries away the pair of electrons as it departs. Frequently, the substrate used in nucleophilic substitution reactions is an alkyl halide (R-X), while the leaving group is a halide anion (X^-). Nucleophilic substitution reactions involve the heterolytic cleavage of the bond between the carbon atom in the substrate and the leaving group. The nucleophile's unshared electron pair establishes a new bond with the carbon atom.

Figure 11. Nucleophilic substitution reaction

$$\text{R-X} + \text{Nu}:^{\ominus} \longrightarrow \text{R-Nu} + \text{X}:^{\ominus}$$

Further investigations were carried out to shed light on the mechanism of nucleophilic substitution reactions and gain insight into the process of configuration inversions, building upon the previous research conducted by Walden. The investigations revealed that nucleophilic substitutions take place through two primary pathways, known as the S_N1 reaction and the S_N2 reaction. Both cases involve the abbreviation "S_N" which stands for substitution and nucleophilic. The significance and distinction between 1 and 2 are discussed in the following two sections.

Substitution: The S_N2 Reaction

Overview and Mechanism of The S_N2 Reaction

An S_N2 reaction occurs in a concerted manner, meaning it happens in a single step without any intermediate species. This reaction occurs when the nucleophile approaches the substrate from a direction that is exactly opposite to the leaving group, with a separation angle of 180°. (Helmreich,

1953) When a nucleophile approaches a molecule, it displaces a leaving group by using its electron pair. The leaving group then exits the molecule from the opposite side, taking the electron pair from the C-X bond with it. During the transition state of the reaction, the formation of the new Nu-C bond and the breaking of the old C-X bond occur simultaneously, while the negative charge is distributed between the incoming nucleophile and the outgoing leaving group. The diagram illustrating the mechanism can be observed in Figure 12 regarding the reaction between OH^- and (*s*)-2-chlorobutane.

Figure 12. The mechanism of S_N2 reaction

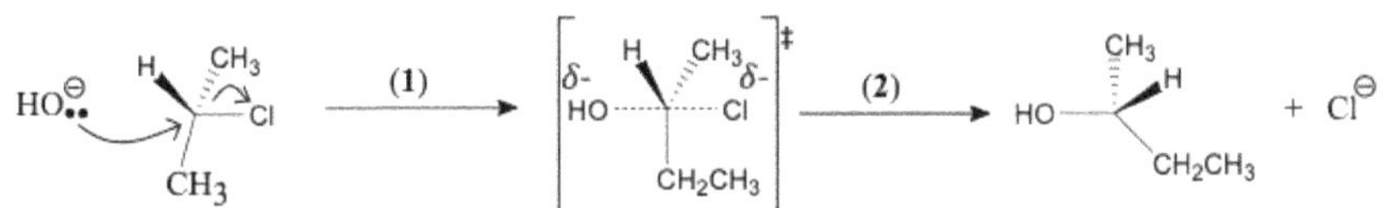

(*S*)-Chlorobutane | Transition state | (*R*)-Butan-2-ol

(1) The nucleophile OH⁻ uses its lone pair electrons to initiate an attack on the carbon atom of the alkyl halide, which is positioned 180° opposite to the departing halogen atom

Transition state with partially formed C-OH bond and partially broken C-Cl bond

(2) The stereochemistry at the carbon atom undergoes inversion as the C-OH bond is fully formed and Cl- departs, taking the electron pair from the former C-Cl bond

Rate of the S_N2 Reaction

The rate of a chemical reaction is directly proportional to the reactants' concentrations. In the 1930s, Hughes and Ingold made the significant finding that certain nucleophilic substitutions follow a first-order reaction rate, meaning that the rate of the reaction is solely determined by the concentration of the alkyl halide and not influenced by the concentration of the nucleophile.(Mortimer & Taylor, 2007) Alternatively, certain nucleophilic substitutions demonstrate a reaction rate that depends on the concentrations of both the alkyl halide and the nucleophile, resulting in a second-order rate. What is the most effective way to elucidate this outcome? The S_N2 reaction between CH_3Cl and OH^- to produce CH_3OH occurs in a single step upon collision and reaction of the substrate and nucleophile (Figure 13).

Figure 13. The S_N2 reaction between CH_3Cl and OH^-

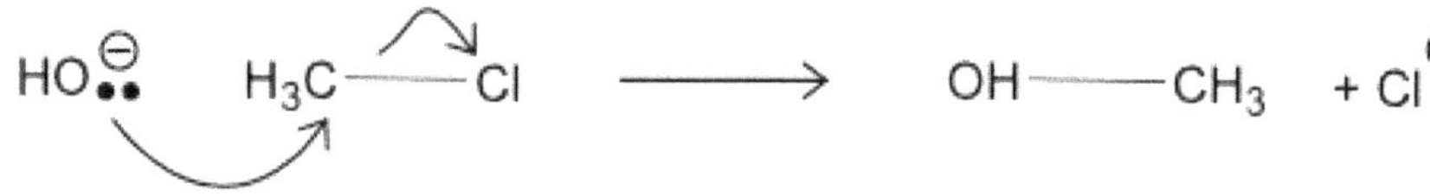

According to the S_N2 mechanism depicted above in Figure 12, the nucleophile approaches the carbon atom that carries the leaving group from the opposite side, known as the back side. The nucleophile's electron pair, located in its highest occupied molecular orbital (HOMO), starts to overlap with the empty orbital (lowest unoccupied molecular orbital or LUMO) of the carbon atom that carries the leaving group. During the course of the reaction, the connection between the nucleophile and the carbon atom becomes stronger, while the bond between the carbon atom and the leaving group becomes weaker. During the reaction, the nucleophile establishes a bond while the leaving group separates. As a result, the substrate undergoes an inversion, causing its tetrahedral bond configuration to be reversed. Nucleophile-carbon bond formation supplies the majority of the energy required to cleave the carbon-leaving group bond.

If there is only one step, then that step is considered to be the rate-determining step. The overall reaction rate is solely determined by the rate of this particular step. According to kinetic theory, the reaction rate is directly proportional to the concentrations of the species involved in the reaction:

Rate of the reaction = $k\,[CH_3Cl][OH^-]$

At a specific concentration of reactants, the reaction occurs at a particular rate. When the concentration of OH^- is doubled, the frequency of collision between the two reactants also doubles, resulting in a corresponding doubling of the reaction rate. Similarly, when the concentration of CH_3Cl is doubled, the reaction rate also doubles. The reaction rate will be directly and proportionally related to both the concentration of CH_3Cl and HO^-. Indeed, it is. Ingold conducted experiments to measure the rates of reactions similar to these and discovered that the rates were directly proportional to the concentration of each reactant, indicating that they followed a second-order reaction kinetics.

Hence, the rationale behind the inclusion of the numeral “2” in S_N2: S_N2 reactions are classified as biomolecular due to their rate being dependent on the concentrations of two substances: alkyl halide and nucleophile. The rate equation is typically expressed in the following form, where k_2 denotes the second-order rate constant.

Rate of the reaction = k_2 $[CH_3Cl][OH^-]$

Stereochemistry of the S_N2 Reaction

Examine the mechanism of the S_N2 reaction depicted in Figure 12 with close attention. The illustration depicts a nucleophile engaging in an attack on the carbon atom situated on the side opposite the leaving group. (McMurry, 2003) Upon close observation of the carbon atom being attacked, it becomes apparent that its substituents undergo an inversion as the reaction progresses, resembling the folding of an umbrella in strong winds. If the carbon atom being targeted is a stereogenic centre, the outcome will be a reversal of its configuration. For instance, the compound (*S*)-2-Chlorobutane undergoes a process called inversion of configuration, resulting in the formation of (*R*)-butan-2-ol. This inversion occurs through a planar transition state (the dotted lines indicate partial bonding) (Figure 14).

Figure 14. The planar arrangement of the transition state for the S_N2 reaction

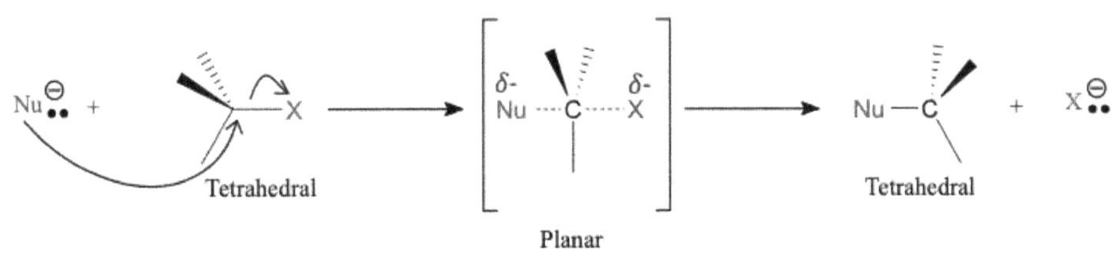

The Nucleophile in the S_N2 Reaction

A proficient nucleophile is indispensable in an S_N2 reaction. To identify a good nucleophile, we must assess its nucleophilicity, which refers to the comparative potency of the nucleophile. There are several factors that can affect nucleophilicity, including charge, basicity, solvent, polarizability, and the properties of the substituents. (Liebman & Greenberg, 1988)

Nucleophiles can exist in either a neutral or negatively charged state. In any given scenario, the nucleophile needs to have a top-notch Lewis base attribute, demonstrating its eagerness to share electrons. The diagram in Figure 15 serves as a helpful reference for the nucleophiles discussed in the section on nucleophilic substitution. After analyzing these two groups of nucleophiles, it becomes clear that an atom with chemical activity, like oxygen, found in a neutral molecule, can also display reactivity when it is part of a negatively charged molecule. As an example, the oxygen atom in the OH^- ion has a negative charge, while the oxygen atom in the H_2O molecule is electrically neutral. Experimental evidence has shown that a nucleophile with a negatively charged reactive atom displays higher reactivity in comparison to a nucleophile with a neutrally charged reactive atom.

Figure 15. Types of nucleophiles

Neutral nucleophiles	Charged nucleophiles
H_2O, NH_3, RNH_2, R_2NH, R_3N, ROH, RCOOH	RO^-, $^-NH_2$, RN^-H, R_2N^-, HS^-, RS^-, RSe^-, Cl^-, Br^-, I^-, F^-, CN^-, ^-OH, RCO_2^-

When moving across a row, nucleophilicity changes. Exploring the fundamental principles of basicity reveals an interesting correlation: as basicity increases from right to left on the periodic table, nucleophilicity also increases. This suggests a strong connection between these two factors. As elements move from left to right on the periodic table, their basicity decreases, which in turn leads to a decrease in their nucleophilicity (Figure 16). When it comes to nucleophilicity, it's important to note that the same principle cannot be applied when comparing halogens within the same group. When it comes to vertical movement within a column, it's important to note that nucleophilicity and basicity don't always go hand in hand. The result can differ depending on the particular solvent used.

Figure 16. Nucleophilicity trend in periodic table

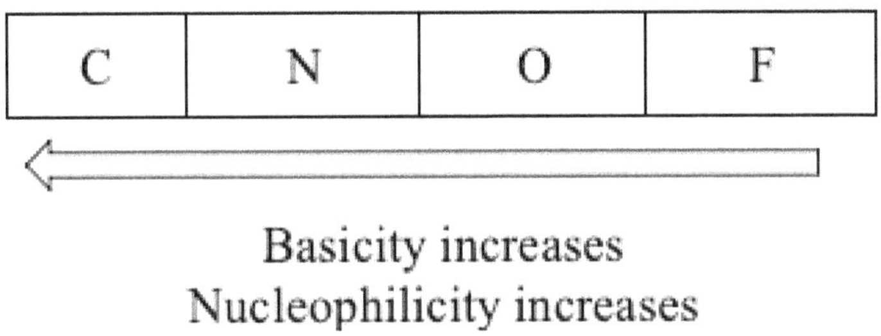

Protic solvents exhibit positive polarization of hydrogen due to the high electronegativity of oxygen and nitrogen atoms, which results from the presence of O-H and N-H bonds. In nucleophilic substitution reactions, protic solvents can enhance the interaction between the positively polarized hydrogen of the solvent molecule and the negatively charged nucleophile. When molecules or ions are surrounded by solvent molecules, they undergo a process known as solvation. When a solute is present, solvent molecules are drawn to it and form associations with the ions. In this case, the solute is an anionic nucleophile. What is the importance of this? When a substance is solvated, its nucleophilic strength is weakened. This means that the solution has the effect of decreasing nucleophilicity. The solvent impedes the nucleophile's capacity to react with an electrophilic carbon by enveloping it in a protective shield. Furthermore, smaller anions undergo more intense solutions due to their higher charge concentration, in contrast to larger anions.

Steric Effect in the S_N2 Reaction

The ability of a nucleophile to approach a substrate and perform an S_N2 reaction is determined by the degree of steric accessibility to the carbon atom that bears the halide group (α–carbon). Substrates that are large and difficult to access, with a carbon atom containing a halide, undergo reactions at a slower rate compared to substrates where the carbon is more easily reached, Figure 17.

Figure 17. Steric hindrance effect to S_N2 reaction

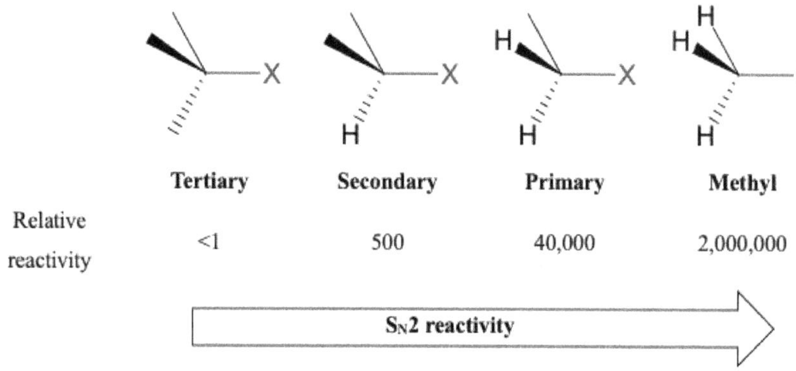

Methyl halides (CH_3-X) exhibit the highest reactivity among substrates, with primary alkyl halides (RCH_2-X) like ethyl or propyl being less reactive. The presence of alkyl branching adjacent to the leaving group significantly retards the reaction rate for secondary halides (R_2CH-X), while additional branching completely inhibits the reaction for tertiary halides (R_3C-X). Vinylic and aryl halides are omitted from this reactivity list due to their inherent inertness towards S_N2 displacements. The rate of S_N2 reaction is influenced by the presence of substitution in the beta position (β), although this impact is less significant compared to the effect observed at the alpha position (α). The presence of two substituents at the beta position leads to a moderate reduction in the rate. Nevertheless, when three substitutions occur at the beta position, the rate of the S_N2 process experiences a substantial decrease, rendering it too sluggish to have any practical utility. Consequently, the S_N2 reaction is typically inhibited when there are three substituents at the alpha position or when there are three substituents at the beta position.

The limited responsiveness is a result of steric hindrance, where the nucleophile attempting to react would need to navigate through a portion of the molecule in order to displace another group. Steric hindrance affects both alkyl halides and nucleophiles. It has been discovered that the S_N2 transition state is highly congested.

Figure 18. Steric hindrance effect to the electrophilic centre in nucleophilic substitution reactions

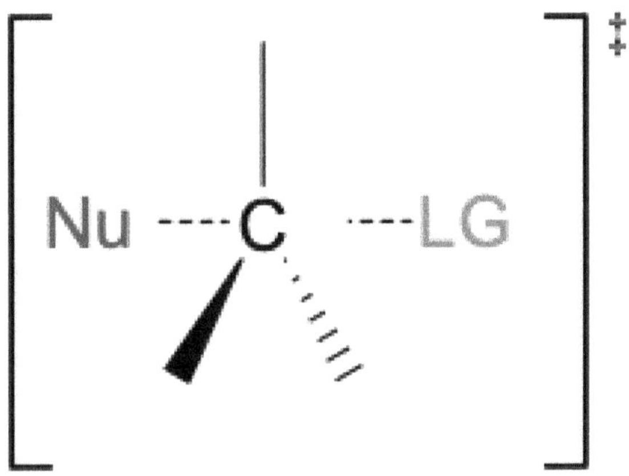

Note that there are a total of 5 groups surrounding the electrophilic center (Figure 18). Sterically hindered nucleophiles exhibit a slower reaction rate compared to nucleophiles that do not have steric bulk.

The Leaving Group in S_N2 Reaction

The nature of the leaving group displaced by the attacking nucleophile is another variable that can influence the S_N2 reaction. This is because the departure of the leaving group plays a crucial role in determining the rate of the reaction. The most favorable leaving groups in S_N2 reactions are those that produce the most stable anions, typically derived from strong acids, as they are expelled with a negative charge. The most prevalent leaving group is a halide ion, specifically I^-, Br^-, or Cl^-, although alternative options are also feasible. Leaving groups such as F^-, OH^-, OR^-, and NH_2^-, which are anions, are infrequently encountered.

Organic Synthesis Refers to the Process of Transforming Functional Groups Utilizing S_N2 Reactions

S_N2 reactions are widely employed in organic synthesis due to their ability to facilitate the conversion of one functional group into another, a process known as functional group transformation or functional group interconversion (Figure 19). The S_N2 reaction depicted in the scheme can convert methyl, primary, and secondary alkyl halides into alcohols, ethers, thiols, thioethers, nitriles, esters, and other compounds. Alkyl chlorides and bromides can be readily transformed into alkyl chlorides through nucleophilic substitution reactions.

On the other hand, stereochemistry is a crucial aspect of S_N2 reactions. S_N2 reactions consistently result in the carbon bearing the leaving group undergoing a configuration inversion. Therefore, by employing S_N2 reactions in our synthesis, we can confidently determine the configuration of our product if we are aware of the configuration of our reactant.

Figure 19. Functional group transformation using S_N2 reaction

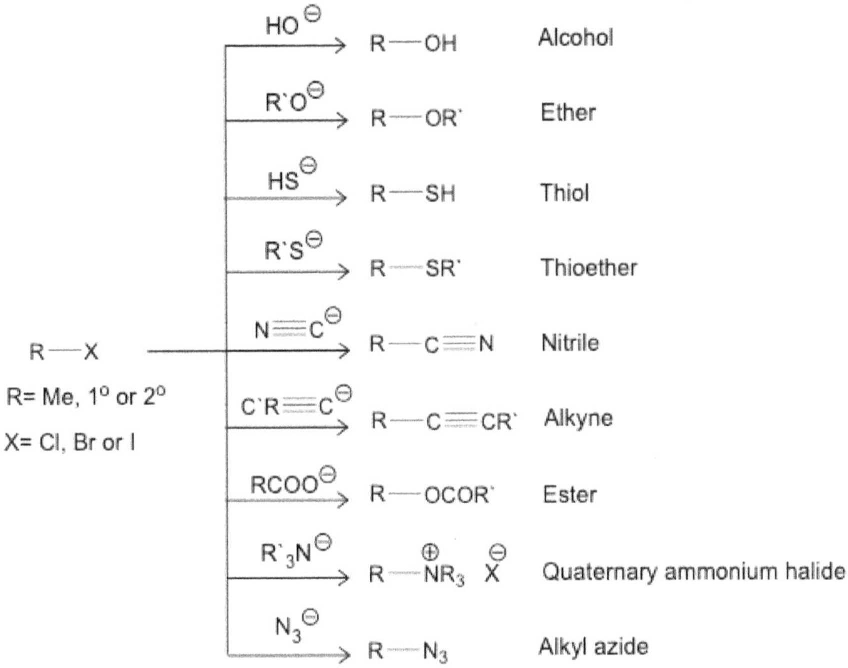

Substitution: The S_N1 Reaction

Overview and Mechanism of The S_N1 Reaction

The majority of nucleophilic substitutions occur through the previously discussed S_N2 pathway. The reaction is more likely to occur when conducted with an unobstructed substrate and a negatively charged nucleophile in a solvent that does not donate protons (aprotic solvent) but is less likely to occur when conducted with an obstructed substrate and a neutral nucleophile in a solvent that donates protons (protic solvent). (Klein, 2017) Consequently, the substitution reaction between a hindered tertiary substrate and neutral, protic water is anticipated to be one of the most sluggish reactions. Surprisingly, the opposite is actually the case. The reaction between the tertiary halide 2-bromo-2-methylpropane $(CH_3)_3CBr$ and H_2O to produce the alcohol 2-methyl-2-propanol occurs at a rate that is over 1 million times greater than the reaction between CH_3Br and H_2O to produce methanol (Figure 20).

Figure 20. Substrate reactivity towards S_N1 reactions

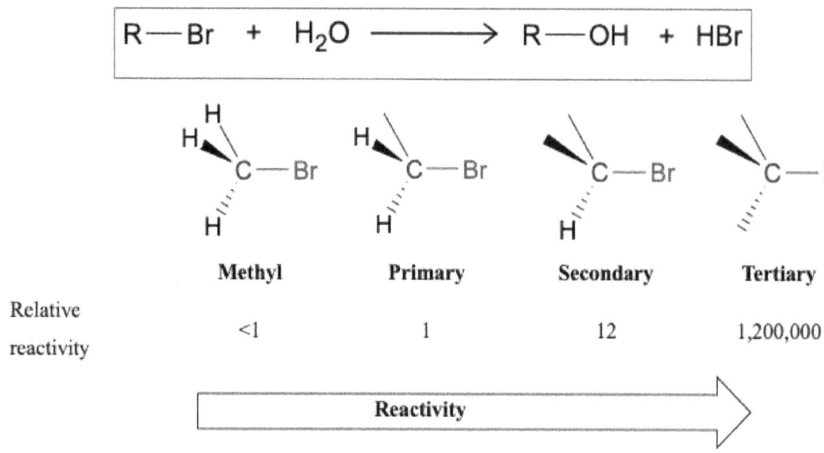

What is the current situation or state of affairs in this place? A nucleophilic substitution reaction is taking place, where a hydroxyl group is displacing a halogen. However, the observed reactivity order appears to be counterintuitive. Based on the discussions we've had, it is evident that the reactions we are observing cannot be explained by the S_N2 mechanism. Consequently, we can infer that they are taking place through an alternative substitution mechanism. The alternative mechanism is referred to as the S_N1 reaction, which stands for substitution, nucleophilic, unimolecular.

In contrast to an S_N2 reaction, where the leaving group is displaced simultaneously with the incoming nucleophile's approach, an S_N1 reaction involves the spontaneous loss of the leaving group before the incoming nucleophile approaches. The departure of the leaving group results in the formation of a carbocation intermediate, which subsequently undergoes a reaction with the nucleophile in a subsequent step to produce the substitution product (Figure 21).

Figure 21. The mechanism of S_N1 reaction

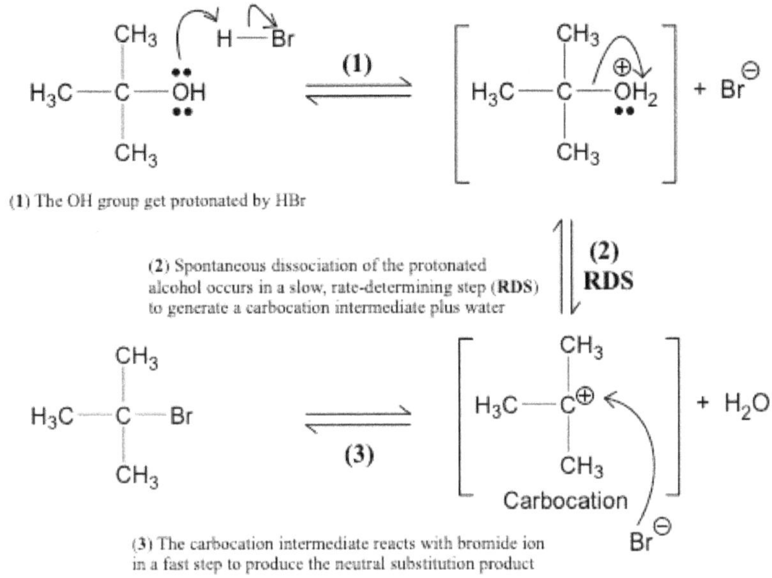

The rapid reactivity of tertiary alcohols towards HBr compared to primary or secondary alcohols can be attributed to a two-step mechanism. This mechanism relies on the formation of stable carbocation intermediates, which are necessary for S_N1 reactions to take place. The rate of the S_N1 reaction is directly proportional to the stability of the carbocation intermediate. Therefore, the order of reactivity of the alcohols with HBr corresponds to the order of stability of the carbocations shown in Figure 22.

Figure 22. Stability order of the carbocations

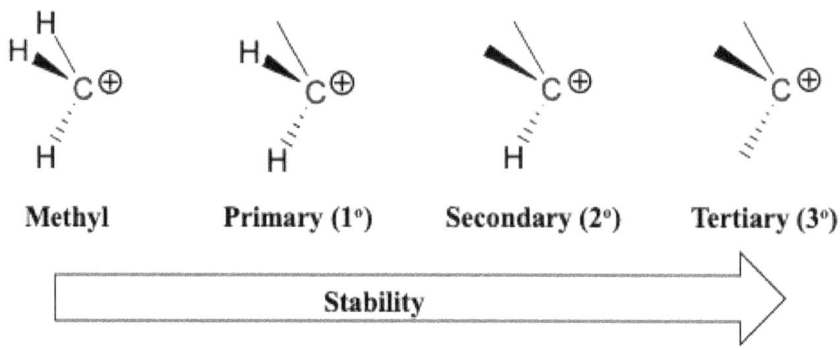

Rate of the S_N1 Reaction

In contrast to an S_N2 reaction, where the rate is influenced by the concentrations of both the substrate and nucleophile, the rate of an S_N1 reaction solely relies on the concentration of the substrate and is unaffected by the nucleophile concentration. Hence, the reason for the "1" in S_N1: S_N1 reactions lies in its unimolecular nature, where the reaction rate is solely determined by the concentration of one substance, namely the substrate. (Mortimer & Taylor, 2007) Put simply, the process can be described as a first-order reaction, where the rate equation does not include the concentration of the nucleophile.

Rate of the reaction = $k \times [RX]$

The concept of S_N1 reactions being unimolecular implies that the substrate undergoes a spontaneous reaction without any participation from the nucleophile. In order to elucidate this outcome, it is imperative to acquire additional information pertaining to kinetics measurements. Multiple organic reactions typically proceed through a series of steps, with one step generally characterized by a transition state of higher energy compared to the others, resulting in a slower rate of reaction. The step with the highest transition-state energy is referred to as the rate-limiting step or rate-determining step. The rate-limiting step in a reaction serves as a bottleneck or traffic jam, preventing any reaction from proceeding faster than this step. In the S_N1 reaction, the absence of nucleophile concentration in the first-order rate equation indicates that it does not play a role in the step that determines the overall rate of the reaction. Instead, it is involved in a different step that does not limit the rate of the reaction. The mechanism depicted in Figure 21 explains these observations.

Stereochemistry of the S_N1 Reaction

Due to the involvement of a carbocation intermediate, the stereochemical outcome of an S_N1 reaction differs from that of an S_N2 reaction. Carbocations are flat, have three sigma bonds formed by the hybridization of three atomic orbitals, and lack chirality. Therefore, when performing an S_N1 reaction on one enantiomer of a chiral reactant and proceeding through an achiral carbocation intermediate, the resulting product will no longer exhibit optical activity. (Solomons et al., 2022)

Specifically, the symmetrical intermediate carbocation has the ability to undergo a reaction with a nucleophile with equal efficiency from both sides (Figure 23). This results in the formation of a racemic mixture, consisting of an equal proportion (50:50) of enantiomers. Put simply, when we perform an S_N1 reaction on a single enantiomer of a chiral substrate and pass through an achiral carbocation intermediate, the molecule briefly loses its chirality, resulting in an optically inactive product.

Figure 23. Stereochemistry of the SN1 reaction

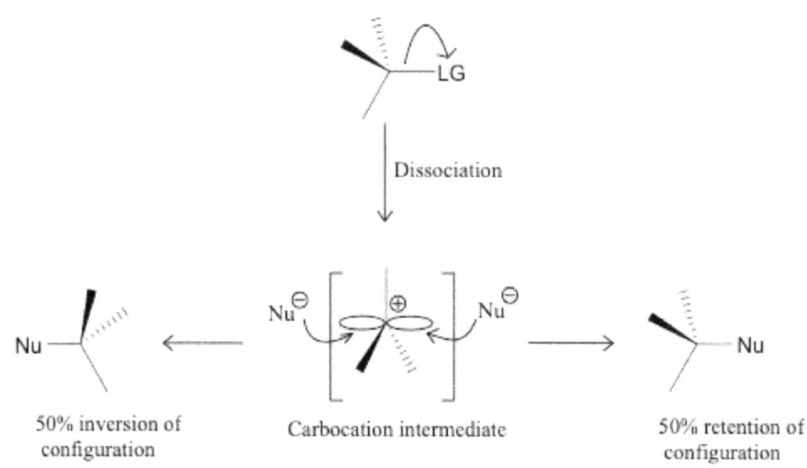

The Nucleophile in the S_N1 Reaction

The nucleophile's nature significantly influences the S_N2 reaction, but it has no impact on an S_N1 reaction. Due to the rate-limiting step in the S_N1 reaction, the nucleophile does not play a role and therefore cannot influence the reaction rate. The reaction between 2-methyl-2-propanol and HX proceeds at a uniform rate, irrespective of whether X represents Cl, Br, or I (Figure 24). Moreover, neutral nucleophiles exhibit comparable effectiveness to negatively charged nucleophiles, thus S_N1 reactions commonly take place in neutral or acidic environments.

Figure 24. The reaction between 2-methyl-2-propanol and HX

Same rate for X= Cl, Br, I

Substrate in the S_N1 Reaction

The Hammond postulate states that any factor which stabilizes a high-energy intermediate also stabilizes the transition state that leads to that intermediate. The rate-limiting step in an S_N1 reaction is the spontaneous and unimolecular dissociation of the substrate, resulting in the formation of a carbocation. This reaction is more likely to occur when a stabilized carbocation intermediate is produced. The rate of the S_N1 reaction is directly proportional to the stability of the carbocation intermediate.

Additionally, it is understood that the order of stability for alkyl carbocations is as follows: tertiary (3°) > secondary (2°) > primary (1°) > methyl. In addition to this list, we should include the resonance-stabilized allyl and benzyl cations. Similar to allylic radicals, allylic and benzylic carbocations exhibit exceptional stability due to the ability of the unpaired electron to spread out over an extended π orbital system. The following scheme demonstrates that an allylic cation possesses two resonance structures (Figure 25). One form of the molecule has a double bond positioned on the "left" side, while the other has a double bond on the "right" side. A benzylic cation exhibits five resonance forms, all of which actively participate in forming the overall resonance hybrid.

Figure 25. Resonance structure of allylic and benzylic carbocations

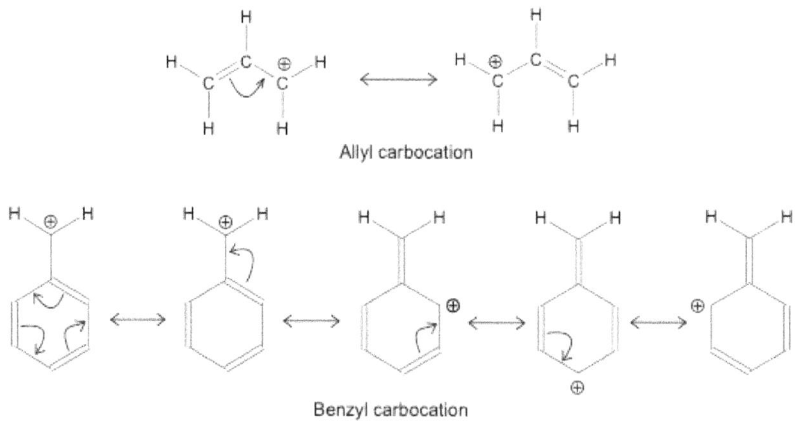

Resonance stabilization results in the primary allylic or benzylic carbocation having a stability comparable to that of a secondary alkyl carbocation, while the secondary allylic or benzylic carbocation exhibits a stability similar to that of a tertiary alkyl carbocation.

Solvolysis

A solvolysis reaction refers to a type of nucleophilic substitution reaction where the nucleophile involved is the solvent molecule itself. The term "solvent + lysis" indicates the cleavage or breaking apart of the reactant molecule by the solvent. The solvolysis reaction is exemplified by the S_N1 reaction of an alkyl halide with water. If the solvent used in the reaction is water, it can be referred to as hydrolysis. If the reaction occurred in methanol, it can be referred to as methanolysis.

The Lack of Reactivity Exhibited by Vinylic and Phenyl Halides

Alkenyl or vinylic halides refer to organic compounds with a halogen atom connected to one carbon atom of a double bond (Figure 26). On the other hand, aryl or phenyl halides are organic compounds with a halogen atom connected to a benzene ring.

Figure 26. Structure of alkenyl and phenyl halides

Alkenyl halide **Phenyl halide**

Alkenyl and phenyl halides typically exhibit low reactivity in S_N1 and S_N2 reactions. Alkenyl and phenyl carbocations are relatively unstable and do not readily form, which is why they do not react in the S_N1 reaction. Alkenyl and phenyl halides are unreactive in S_N2 reactions due to the stronger halogen-carbon bond compared to alkyl halides. Additionally, the presence of a double bond or benzene ring causes the electrons to repel the approach of a nucleophile from the back side.

Nucleophilic Substitutions on Alcohols: Strategies for Inducing Departure of the Hydroxyl (OH) Group

Alcohols do not undergo reactions with nucleophiles. Put simply, OH^- is always incapable of acting as a leaving group. What is the reason for not doing so? To begin with, the hydroxide ion is highly alkaline. If the nucleophile were sufficiently potent to displace the hydroxide ion, it would be more than capable of extracting the proton from the alcohol (Figure 27). (Bruice, 2015)

Figure 27. Nucleophilic reaction on alcohols

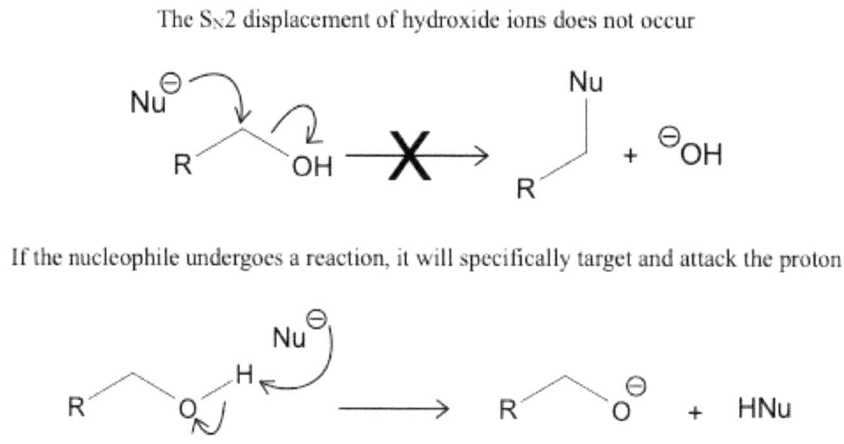

However, alcohols are desirable for nucleophilic substitution reactions due to their facile synthesis. An effective approach is to protonate the hydroxyl group using a potent acid (Figure 28). This will be effective only if the nucleophile is compatible with a potent acid, although numerous nucleophiles meet this criterion. An excellent illustration is the production of *t*-BuCl from *t*-BuOH through agitation with concentrated HCl. The reaction at hand is clearly an S_N1 reaction, with the *t*-butyl cation serving as an intermediate.

Figure 28. Nucleophilic substitution of tertiary alcohol with a strong acid

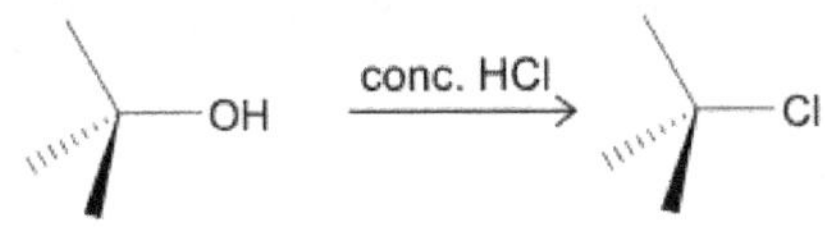

The mechanism of the reaction:

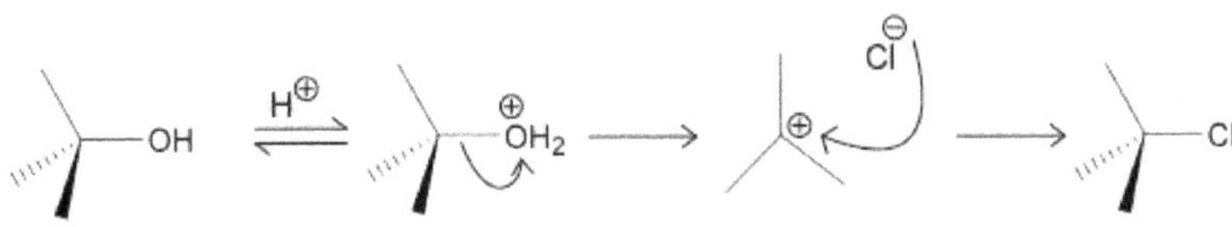

Analogous techniques can be employed to synthesize both secondary and primary alkyl halides. The reactions must proceed through the S_N2 pathway, involving the substitution of the protonated hydroxyl group with a halide. An alternative method for substituting OH involves enhancing its leaving group properties by combining it with an element that forms highly stable bonds with oxygen (Figure 29). Phosphorus and sulfur are the most favored options. The synthesis of primary alkyl bromides using PBr_3 is typically effective. The phosphorus reagent undergoes an S_N2 reaction with the OH group, resulting in the displacement of an oxygenation bonded to phosphorus. This reaction is favorable due to the stabilization of the anion by phosphorus.

Figure 29. Substitution of the hydroxyl group by combination with an element

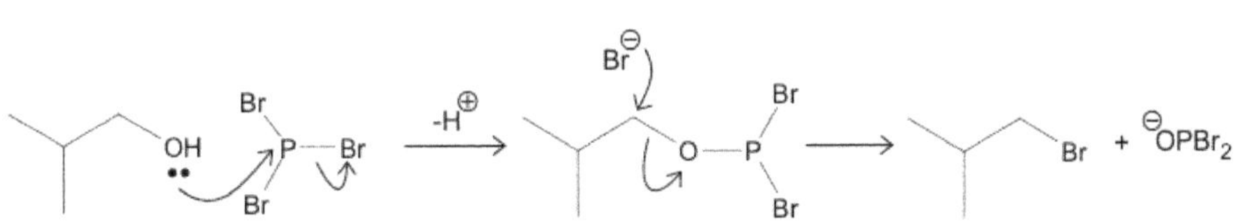

One commonly employed method to convert a hydroxyl group into an effective leaving group is by transforming it into a sulfonate ester. Primary and secondary alcohols can be readily transformed into sulfonate esters through the reaction with sulfonyl chlorides and a base. Sulfonate esters are commonly found in a crystalline form and are extensively utilized, to the point where they have been assigned informal names such as tosylates for p-toluenesulfonates and mesylates for methanesulfonates. Additionally, the functional groups in these esters have been designated with the 'organic element' symbols Ts and Ms (Figure 30).

Figure 30. Structure of tosylate and mesylate

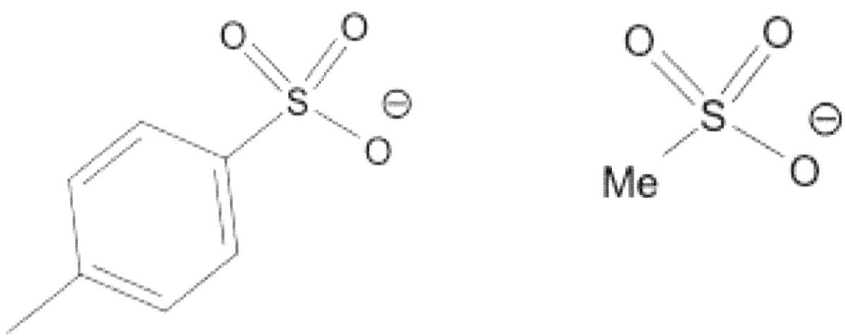

Tosylate (Ts) **Mesylate (Ms)**

Instead of employing a two-step process involving the conversion of the OH group to a sulfonate ester followed by displacement, it is feasible to utilize a method that enables the direct addition of an alcohol to a reaction mixture, resulting in the formation of an S_N2 product in a single operation. The following is an explanation of the Mitsunobu reaction. In this chemical reaction, the alcohol molecule acts as the electrophile, while the nucleophile is typically a relatively weak compound, such as the conjugate base of a carboxylic acid. Additionally, there are two other reagents involved. One of the reagents is triphenylphosphine, also known as Ph_3P. Phosphines exhibit nucleophilic properties, but they do not possess basic characteristics similar to amines. Further analysis should be dedicated to the other reagent. The compound is formally known as diethyl azodicarboxylate, but it is commonly referred to as DEAD.

In the initial phase of the Mitsunobu reaction (Figure 31), neither the alcohol nor the additional nucleophile are involved. The phosphine compound reacts with the relatively weak N=N double bond, resulting in the formation of an anion that is stabilized by one of the ester groups.

Figure 31. Stage one of the Mitsunobu reaction

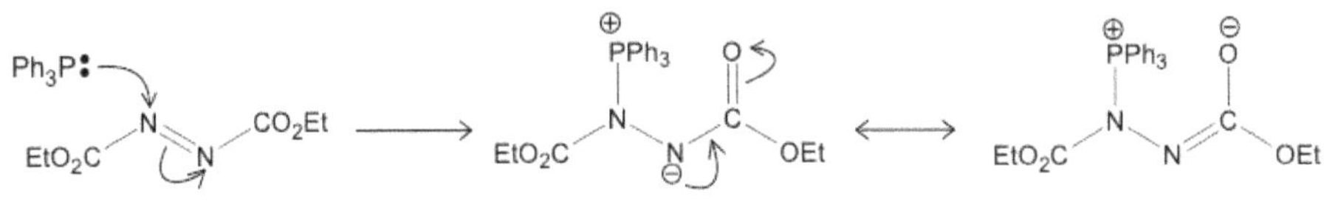

It is important to observe that the nucleophile has been introduced in the form of its conjugate acid 'HNu'. In many cases, this conjugate acid can be a carboxylic acid, such as benzoic acid. The anion generated in the initial step is sufficiently alkaline to extract a proton from the acid, resulting in the formation of Nu^-, which is then available for further reaction (Figure 32).

Figure 32. Stage two of the Mitsunobu reaction

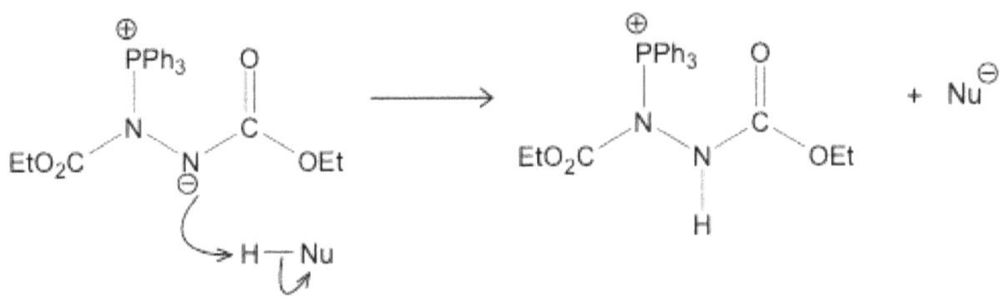

Oxygen and phosphorus exhibit a significant attraction, leading to the alcohol attacking the positively charged phosphorus (Figure 33). This results in the displacement of a second nitrogen anion in a phosphorus-centered S_N2 reaction. The nitrogen anion formed in this step is stabilized through conjugation with the ester, but quickly abstracts the proton from the alcohol, resulting in the formation of an electrophilic R–O–PPh_3^+ species and a by-product, which is the reduced form of DEAD.

Figure 33. Stage three of the Mitsunobu reaction

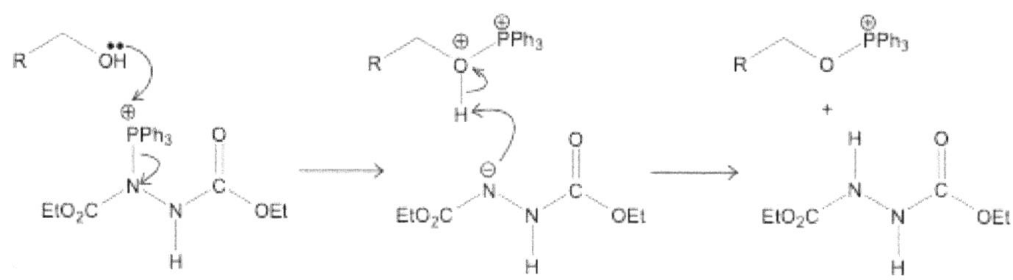

Ultimately, the negatively charged form of the nucleophile is able to initiate an attack on the phosphorus derivative of the alcohol in a typical S_N2 reaction, targeting the carbon atom where the phosphine oxide group will be expelled (Figure 34). We have reached the merchandise.

Figure 34. Stage four of the Mitsunobu reaction

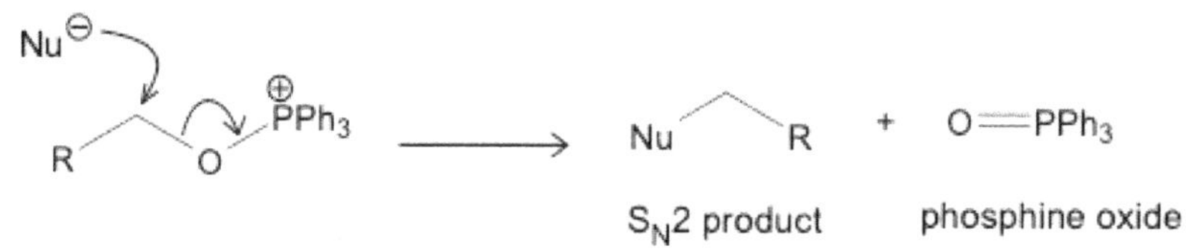

The entire process occurs in a single operation. All four reagents are combined in a single flask, resulting in the formation of three products: phosphine oxide, a reduced azo diester with two NH bonds replacing the N=N double bond, and the product of an S_N2 reaction on the alcohol. An alternative perspective on this reaction is that it involves the formal loss of a water molecule, where the hydroxyl group (OH) is removed from the alcohol and a hydrogen atom (H) is removed from the nucleophile. The atoms ultimately form highly stable molecules, with the P=O and N–H bonds exhibiting strength, while the N=N bond is comparatively weak.

This compensates for the loss of the strong C–O bond present in the initial alcohol compound.

If all the information provided is accurate, then the crucial S_N2 step should result in inversion, as is typically observed in S_N2 reactions. An advantageous aspect of the Mitsunobu reaction is its ability to consistently substitute an OH group with a nucleophile while inverting the configuration. An exemplary instance of this phenomenon is the formation of esters from secondary alcohols through inversion. In normal ester formation, the C-O bond of the alcohol remains intact, resulting in its retention. By comparing these two reactions, one can observe the location of the circled oxygen (and hydrogen) atoms (Figure 35).

Figure 35. The formation of ester by the Mitsunobu reaction to the normal method

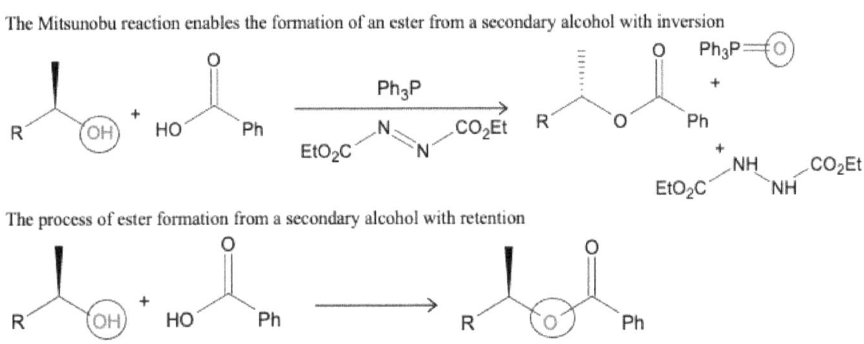

How Can We Determine the Applicable Mechanism (S_N1 or S_N2) for a Specific Organic Compound?

Substitution reactions at saturated carbon atoms occur through one of two distinct mechanisms, each exhibiting a contrasting reliance on the characteristics of the nucleophile. It is crucial to have the ability to anticipate the mechanism that is most likely to occur in a given reaction. Instead of conducting kinetic experiments to determine this, we can provide you with a few straightforward guidelines to predict which mechanism will be in operation. (Clayden et al., 2012) The variables that influence the process of the reaction also contribute to the understanding of why that process occurs (Table 1).

Table 1. Overview reactivity of S_N1 and S_N2 reactions

Type of structure	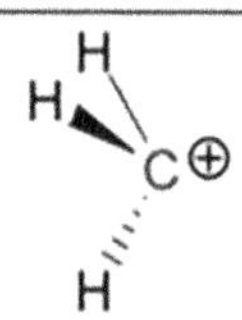 **Methyl**	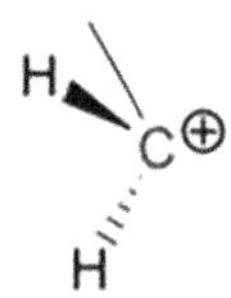 **Primary (1°)**	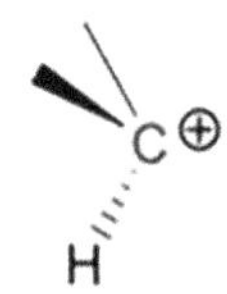 **Secondary (2°)**	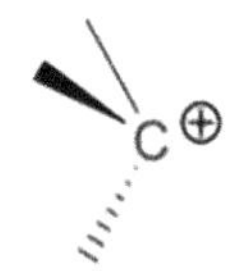 **Tertiary (3°)**
Type of substitution	S_N2	S_N2	S_N2/ S_N1	S_N1
• In Secondary substrate; S_N2 substitution occurs if a weakly basic nucleophile is used in a polar aprotic solvent				
• In Secondary substrate; S_N1 substitution occurs if a weakly basic nucleophile is used in a protic solvent				

The carbon skeleton's structure is the most crucial factor. An insightful generalization is that compounds capable of forming stable carbocations tend to undergo reactions via the S_N1 mechanism, whereas compounds lacking this ability are compelled to react via the S_N2 mechanism. The stability of carbocations is directly proportional to the number of substituents they possess. Therefore, carbocations with a higher number of carbon substituents at the reaction centre are more likely to undergo reactions through the S_N1 mechanism. The stability of cations, which is determined by structural factors, often results in slower S_N2 reactions. Highly substituted compounds exhibit favorable reactivity in S_N1 reactions but exhibit unfavorable reactivity in S_N2 reactions due to the hindered access of the nucleophile to the reaction centre caused by the presence of substituents. The S_N2 reaction is more favorable when the reaction centre contains only hydrogen atoms. Methyl groups exhibit the highest reactivity in S_N2 reactions. The impacts of the most basic structural modifications are outlined in Table 1 (where R represents a straightforward alkyl group such as methyl or ethyl). The sole ambiguous scenario pertains to the secondary alkyl derivative, which has the capacity to undergo reaction *via* either mechanism, albeit with limited proficiency in both cases.

ELIMINATION REACTIONS

So far, we have only examined substitution reactions. However, it is important to note that there are actually two types of reactions that can occur when a nucleophile/Lewis base reacts with an alkyl halide. The nucleophile has the ability to either replace the halide by reacting with carbon in an S_N1 or S_N2 reaction or induce the elimination of HX by reacting with a nearby hydrogen, leading to the formation of an alkene (Figure 36) (Team, 1973). These reactions exhibit similarities and frequently engage in competitive interactions.

Figure 36. Substitution and elimination reactions

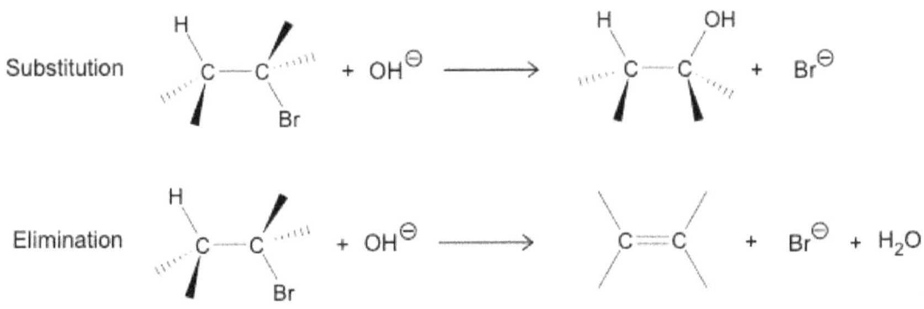

The Impact of the Nucleophile on the Choice Between Elimination and Substitution Reactions

In Figure 36, you can see molecules with leaving groups being attacked at two different electrophilic sites: the carbon to which the leaving group is attached and the hydrogen atoms on the carbon next to the leaving group. Attacking carbon causes substitution while attacking hydrogen causes elimination. Since strong bases attack protons, it is generally true that the more basic the nucleophile, the more likely elimination will replace substitution as the primary reaction of an alkyl halide.

Basicity

The potency of the base is a determining factor in the type of reaction that takes place. Here is an illustration of this concept in action: Substitution occurs when a weak base, such as ethanol (EtOH), is used (Figure 37-left), whereas elimination occurs when a strong base, such as ethoxide ion, is used (Figure 37-right).

Figure 37. The impact of the base strength on the reaction pathway

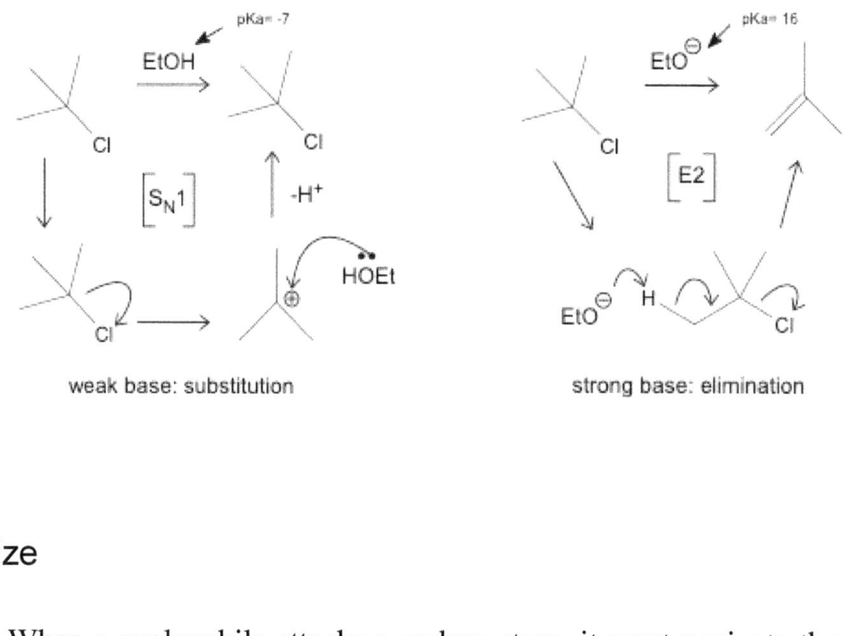

Size

When a nucleophile attacks a carbon atom, it must navigate through the substituents surrounding the carbon atom. Even in the case of primary alkyl halides without any hindrance, there is still one alkyl group attached to the carbon atom. This is a contributing factor to the sluggishness of S_N2 reactions on hindered alkyl halides, as the nucleophile encounters challenges in accessing the reactive center (Figure 38). Accessing a hydrogen atom that is more readily available in an elimination reaction is significantly simpler. Consequently, when employing large basic nucleophiles, elimination becomes the favored reaction pathway over substitution, even for primary alkyl halides. Potassium t-but-oxide is an excellent catalyst for facilitating elimination reactions and preventing substitution reactions. The presence of a bulky alkyl substituent hinders the nucleophilic attack of the negatively charged oxygen on carbon in a substitution reaction, while it readily attacks hydrogen.

Figure 38 The impact of the nucleophile's size on the reaction pathway

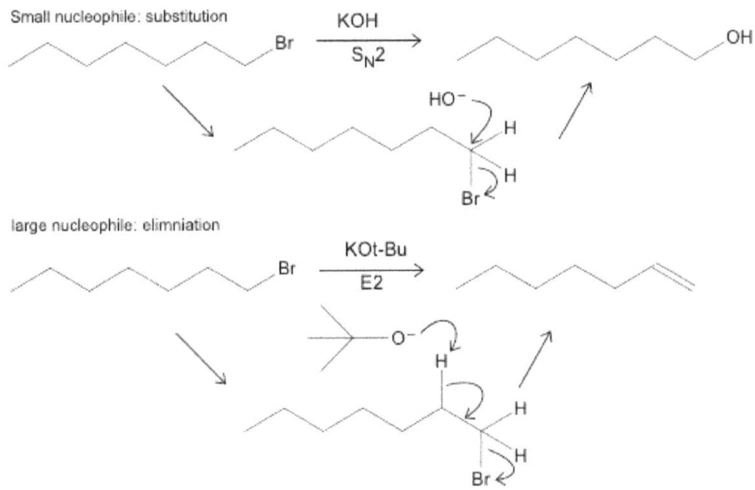

Temperature

Temperature plays a crucial role in determining whether a reaction is an elimination or a substitution. During an elimination reaction, the number of molecules increases from two to three (please take note of this count). A substitution reaction involves the formation of two new molecules from two initial molecules. The two reactions differ in the change in entropy during the reaction. The change in entropy, ΔS, is greater for elimination compared to substitution.

The equation $\Delta G = \Delta H - T\Delta S$ represents the relationship between the change in Gibbs free energy (ΔG), the change in enthalpy (ΔH), temperature (T), and the change in entropy. The equation above states that a reaction with a positive change in entropy (ΔS) becomes more advantageous (ΔG becomes more negative) at higher temperatures. Eliminations should be preferred at high temperatures, and this is indeed the situation: the majority of eliminations you will observe are carried out at room temperature or higher.

Zaitsev's Rule

Elimination reactions are more intricate than substitution reactions due to various factors. One issue that arises is the problem of regiochemistry. What product is formed when an unsymmetrical halide loses HX? Elimination reactions typically yield a combination of alkene products, and our usual goal is to determine the predominant product.

Zaitsev's rule, established in 1875 by Russian chemist Alexander Zaitsev, states that base-induced elimination reactions typically result in the formation of a more stable alkene product. (Walker, 2018) This means that the alkene with a greater number of alkyl substituents on the carbon atoms of the double bond is usually favored, although there may be exceptions to this rule. In the two cases that follow, the alkene product with a higher degree of substitution is the more prevalent one (Figure 39).

Figure 39. Elimination products as predicted by Zaitsev's rule

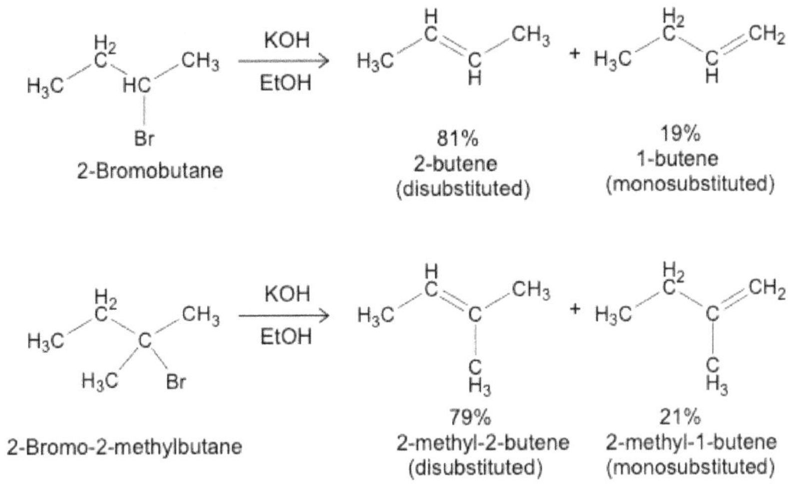

Through the utilization of the potent base hydroxide as shown in Figure 39, we steer these reactions towards elimination as opposed to substitution. Both scenarios involve two distinct sets of neighboring hydrogen atoms that are accessible for the elimination reaction. If the rate of each potential elimination was uniform, we would anticipate that the quantities of the isomeric elimination products would correspond to the number of hydrogens capable of participating in that reaction. Given that there are three primary hydrogens and two secondary hydrogens on the beta-carbons in 2-bromobutane, it can be inferred from statistical analysis that the products will have a ratio of 1-butene to 2-butene of 3:2. This phenomenon is not observed, and the latter prevails by a ratio of 4 to 1. The results indicate a strong preference for the double bond in the more highly substituted product, as predicted by Zaitsev's Rule.

Another complicating factor in the study of elimination reactions is the occurrence of multiple mechanisms, similar to substitutions. In the next couple of sections, we will examine three prevalent mechanisms, namely the E1, E2, and E1cB reactions, which vary in terms of the timing of C–H and C–X bond cleavage.

The E2 Reaction

E2 reactions commonly occur with secondary and tertiary alkyl halides, while a hindered base is required for primary halides. The occurrence of this process is facilitated by a single-step concerted reaction with a single transition state. The rate of this mechanism follows second-order kinetics and is determined by both the base and alkyl halide. An effective leaving group is necessary as it plays a crucial role in the rate-determining step. In order to form a *pi* bond, it is necessary for the leaving groups to be in the same plane. Additionally, the carbon atoms undergo a change in hybridization states from sp^3 to sp^2.

Similar to the S_N2 reaction, the E2 reaction occurs in a single step without any intermediate stages (Figure 40). The base in this reaction is represented by B, while the leaving group, usually a halogen, is represented by X. There exists a single transition state that demonstrates the simultaneous reaction of the base attracting the hydrogen and the halogen accepting the electrons from the bond. The product can exhibit both eclipsed and staggered conformations depending on the transition states. Eclipsed products exhibit synperiplanar transition states, whereas staggered products display antiperiplanar transition states. The staggered conformation is typically the predominant product due to its lower energy conformation.

Figure 40. The mechanism of E2 reaction

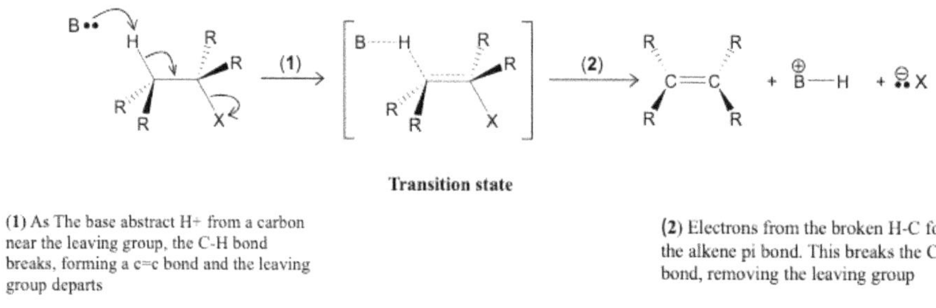

The S_N2 and E2 reactions exhibit several common characteristics. Both reactions necessitate the presence of effective leaving groups, and both mechanisms occur in a concerted manner. S_N2 reactions necessitate a nucleophile with high reactivity, while E2 reactions necessitate a base with significant strength. Nevertheless, a proficient nucleophile frequently exhibits strong basicity. Due to the similarity in conditions between the two reactions, they frequently engage in competition with one another. The outcome of the competition is influenced by three factors: the existence of antiperiplanar β-hydrogens, the level of α and β branching, and the relative nucleophilicity and basicity of the reactant species.

For an E2 elimination to take place, the presence of antiperiplanar β-hydrogens is necessary. If there are no other reactions present, the S_N2 reaction will be the most prevalent. Similarly, the S_N2 nucleophile requires an unobstructed pathway to the σ* C-LG antibond. The presence of α and β branching obstructs this pathway and decreases the ratio of S_N2 to E2 reactions. E2 reactions can still occur even when there are many branches in the molecule. This is because E2 reactions rely on the presence of β-hydrogens, which are more easily accessible compared to the antibonding σ* C-LG orbital.

The preferred mechanism is determined by the identity of the nucleophile or base. E2 reactions necessitate the use of potent bases. S_N2 reactions necessitate the presence of highly effective nucleophiles. Consequently, a nucleophile with good reactivity but low basicity will promote S_N2 reactions, whereas a nucleophile with weak reactivity but high basicity will promote E2 reactions. Large nucleophiles face difficulties in accessing the α-carbon, resulting in an increased ratio of E2 to S_N2 reactions. Polar, aprotic solvents enhance nucleophilicity, consequently accelerating the S_N2 reaction rate.

The E1 Reaction

The E2 reaction is similar to the S_N2 reaction, and similarly, the S_N1 reaction has a corresponding reaction known as the E1 reaction (which stands for elimination, unimolecular). The E1 reaction can be represented as depicted in Figure 41, involving the removal of HCl from 2-chloro-2-methylpropane.

E1 eliminations commence with the identical unimolecular dissociation observed in the S_N1 reaction, resulting in the formation of carbocation. However, in E1 eliminations, the dissociation is subsequently followed by the removal of H^+ from the neighboring carbon, rather than by substitution.

Figure 41. The mechanism of E1 reaction

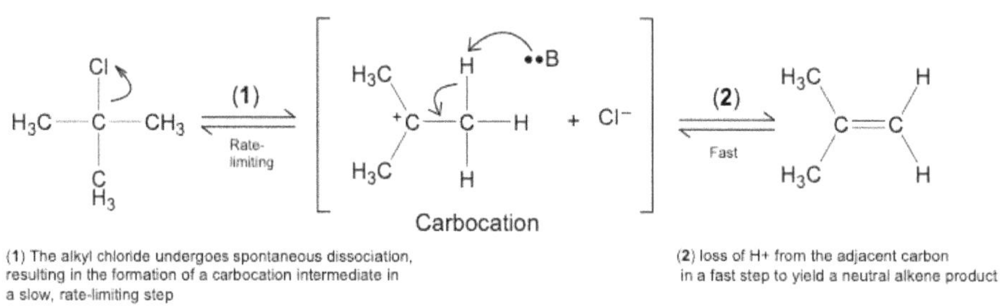

S_N1 and E1 reactions are often interconnected and challenging to separate due to their shared characteristics and the fact that the conditions that promote one reaction also promote the other. We have previously acquired knowledge about the distinctive attributes of S_N1 reactions and the determinants that promote their occurrence. This concept can also be applied to E1 reactions. The S_N1 and E1 mechanisms exhibit unique characteristics. These processes involve multiple steps, occur when the step that determines the rate forms carbocation intermediates, and exhibit kinetics that mimic first-order (unimolecular) reactions. Specifically, the rate is determined by multiplying the rate constant by the substrate concentration.

The following factors promote the S_N1 and E1 mechanisms: use of substrates with steric hindrance, use of weak (neutral) and small nucleophiles, heating, and use of solvents with moderate to high polarity. The S_N1 mechanism results in the formation of substitution products, while the E1 mechanism leads to the formation of alkenes. Substrates that are susceptible to S_N1 reactions are also susceptible to E1 reactions. There are two primary categories: alkyl halides with secondary and tertiary carbon atoms, and alcohols with secondary and tertiary carbon atoms.

The E1cb Reaction

The E1cB reaction differs from the E1 reaction in that it proceeds *via* a carbanion intermediate instead of a carbocation intermediate (Figure 42). The proton is abstracted by a base in a slow step that limits the rate. This results in the formation of an anion, which then removes a leaving group from the carbon atom next to it. The reaction is frequently observed in substrates with a deficient leaving group, such as $^{-}$OH, located two carbons away from a carbonyl group, as in HOC–CH–C=O. The presence of a poor leaving group discourages the occurrence of alternative E1 and E2 reactions. Additionally, the adjacent hydrogen becomes unusually acidic due to the resonance stabilization of the anion intermediate caused by the carbonyl group.

Figure 42. The mechanism of E1cB reaction

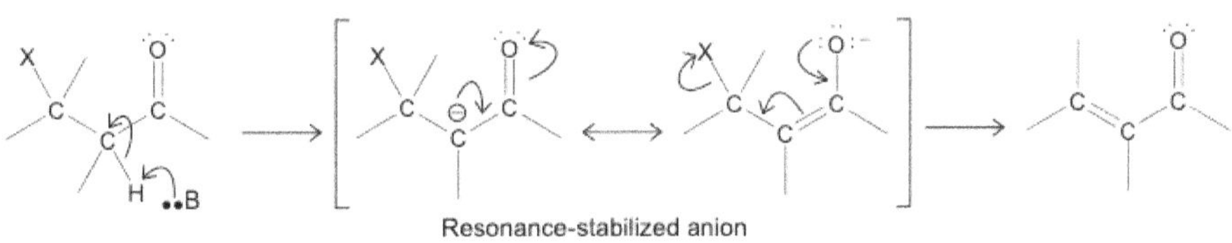

AN OVERVIEW OF REACTIVITY: SN1, SN2, E1, E1CB, AND E2

After observing five distinct types of nucleophilic substitution/elimination reactions, you might be curious about how to anticipate the outcome in any given scenario. Which process, substitution or elimination, will take place? Will the reaction exhibit a unimolecular or bimolecular nature? There is no definitive answer to these questions, but it is possible to identify certain patterns and draw some general conclusions, (Clayden et al., 2012) table 2 provides a concise overview of the anticipated reactivity patterns associated with different structural classes of alkyl halides.

Table 2. Overview of reactivity:S_N1, S_N2, E1, E1cB, E2

Methyl halide	**• An S_N2 substitution reaction occurs irrespective of the quality of the nucleophile employed.**
Primary halide	• S_N2 substitution takes place when a strong nucleophile is utilised, such as I^-, Br^-, RS^-, NH_3, or CN^-. • E2 elimination takes place when a powerful base with steric hindrance, such as DBU and t-BuO^-, is utilised. • E1cB elimination takes place when the departing group is positioned two carbons away from a carbonyl group.
Secondary halide	• S_N2 substitution takes place when a nucleophile with weak basicity is employed in a polar aprotic solvent. • The use of a strong base leads to the predominance of E2 elimination. • E1cB elimination occurs when the departing group is located two carbons away from a carbonyl group. • If a weakly basic nucleophile is used in a protic solvent, secondary allylic and benzylic alkyl halides can also undergo S_N1 and E1 reactions.
Tertiary halide	• E2 elimination takes place when a base is employed. • S_N1 substitution and E1 elimination reactions take place simultaneously under neutral conditions, such as in pure ethanol or water. • E1cB elimination occurs when the departing group is located two carbons away from a carbonyl group.

Regarding Table 2, it is important to note the following:

- The elimination of methyl halides is not possible due to the absence of protons in suitable positions.
- A higher degree of branching promotes elimination reactions rather than substitution reactions. Additionally, highly basic hindered nucleophiles tend to undergo elimination reactions unless no other reaction pathway is available.
- Effective nucleophiles undergo substitution through the S_N2 mechanism unless the substrate is tertiary. In that case, the intermediate cation can undergo elimination through the E1 mechanism in addition to substitution through S_N1.
- Elevated temperatures promote elimination reactions by enhancing the significance of entropy in the free energy change of the reaction

($\Delta G = \Delta H - T\Delta S$). This method effectively guarantees E1 in situations where there is ambiguity.

SUBSTITUTION AND ELIMINATION IN BIOLOGICAL PROCESSES

Chemistry, regardless of whether it is conducted in laboratory flasks by chemists or within cells by living organisms, adheres to identical principles. Biological reactions exhibit the same addition, elimination, substitution, and rearrangement mechanisms as laboratory reactions. S_N1 and S_N2 reactions are frequently observed in biological chemistry, specifically in the biosynthetic pathways of terpenoids, which are numerous plant-derived substances. Contrary to the usual occurrence in the laboratory, halogens are not commonly found in biological molecules, except for some marine organisms. Phosphates, water, alcohols, and thiols are frequently encountered leaving groups in biochemical reactions. Often, when a bond is broken, the leaving group is protonated by an acidic group on the enzyme. As an illustration, the hydroxyl group of an alcohol is initially transformed through enzymatic means into a phosphate ester to enhance its ability to depart from a molecule. This phosphate ester can exist as a basic monophosphate **A**, a diphosphate **B**, or a nucleotide monophosphate **C** (Figure 43). The outcome phosphates are highly effective leaving groups because of the resonance delocalization of the negative charge they acquire during the reaction.

Figure 43. Alcohol transformation to phosphate ester

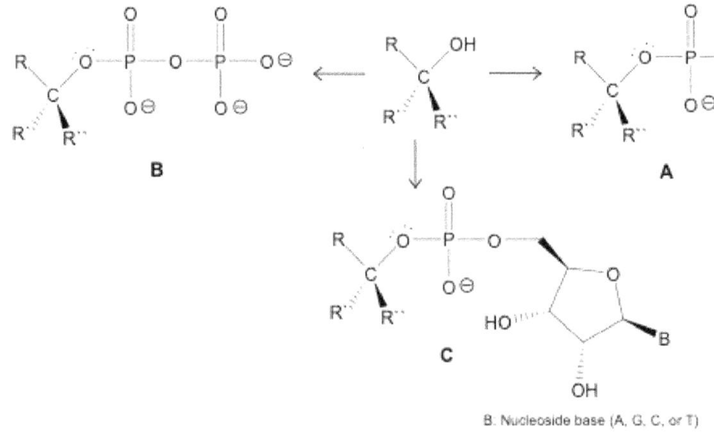

For instance, the S_N1 reaction takes place in the process of synthesizing DNA nucleotides (Figure 44). The hydroxyl group on ribofuranose is transformed into a diphosphate, which is a more effective leaving group. Ammonia acts as the nucleophile in the second step of this S_N1-like reaction.

Figure 44. S_N1 reaction in biological synthesis of DNA nucleotides

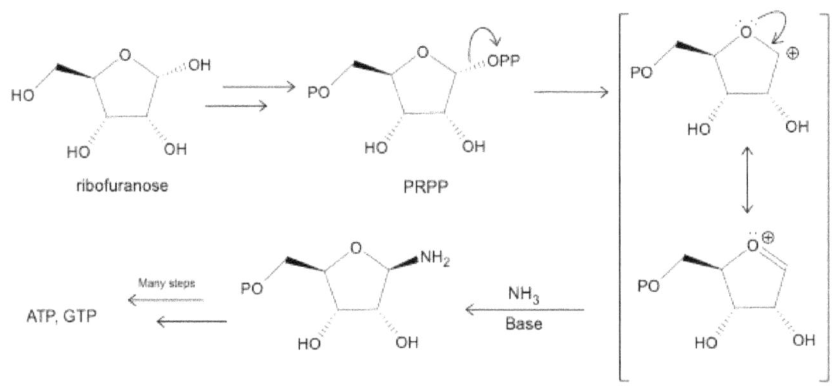

S_N2 reactions in biochemistry are significantly exemplified by methyltransferase enzymes that are catalyzed by *S*-adenosyl methionine (SAM). In an S_N2 reaction, the methyl group is transferred from *S*-adenosylmethionine (SAM) to the amine group on the nucleotide base adenosine (Figure 45).

Figure 45. S_N2 reaction in the biological process

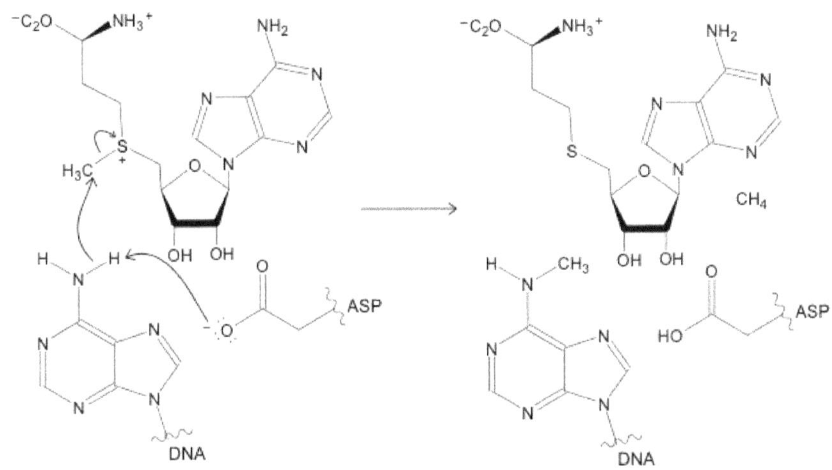

Eliminations, similar to substitutions, are common events in biological pathways. An example of a biological E1-like elimination step can be found in the histidine biosynthetic pathway (Figure 47). Specifically, we are examining the first step in the reaction, which involves the formation of an enol. The second step is a simple tautomerization from the enol to the ketone product.

Figure 46. E1 reaction in histidine biosynthesis pathway

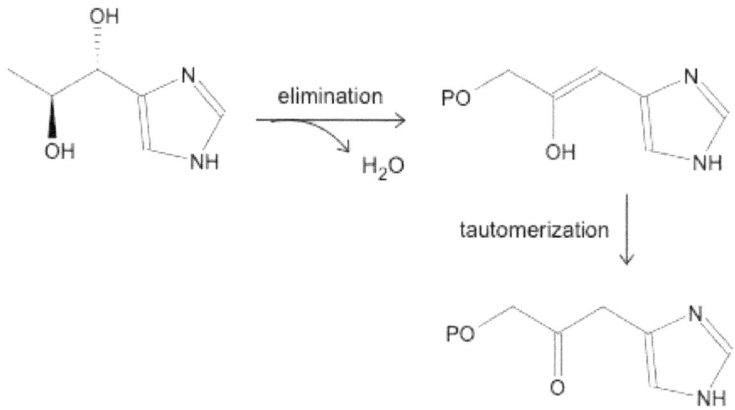

The E1cB mechanism is the predominant form of elimination observed in biological reactions. The substrate typically consists of an alcohol and the hydrogen atom that is eliminated is usually located next to a carbonyl group, similar to reactions conducted in a laboratory setting. An illustrative instance arises in the process of fat biosynthesis (Figure 47), wherein a 3-hydroxybutyryl thioester undergoes dehydration to form the corresponding unsaturated thioester. The enzyme contains an amine functional group as the base of the reaction.

Figure 47. E1cB reaction in fat biosynthesis

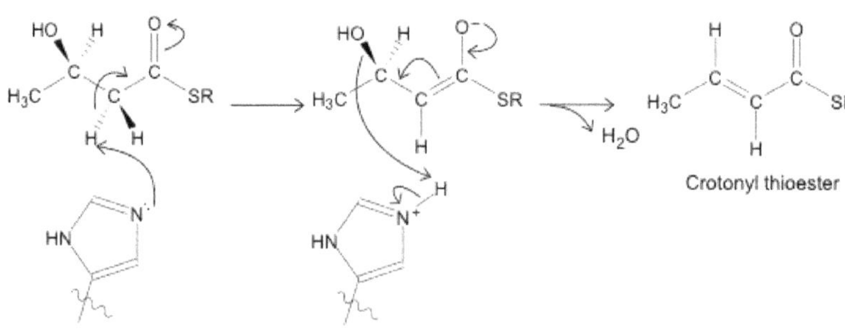

SUMMARY OF THE CHAPTER

Alkyl halides are infrequently present in terrestrial organisms, but the reactions they undergo are highly significant and extensively studied in organic chemistry. This chapter focused on the nomenclature and preparation of alkyl halides, as well as providing an in-depth analysis of their substitution and elimination reactions.

Alkyl halides are typically synthesized from alcohols through the use of HX (for tertiary alcohols) or $SOCl_2$ or PBr_3 (for primary and secondary alcohols). Alkyl halides undergo a reaction with magnesium metal to produce organomagnesium halides, which are commonly known as Grignard reagents (RMgX). Grignard reagents, possessing both nucleophilic and basic properties, undergo reactions with acids to produce hydrocarbons.

When an alkyl halide reacts with a nucleophile or base, it can undergo either a substitution or elimination reaction. Nucleophilic substitution reactions proceed through two distinct mechanisms: S_N2 and S_N1. In the S_N2 reaction, the nucleophile approaches the substrate from a direction opposite to the leaving group, causing a complete inversion of configuration at the carbon atom, resembling an umbrella-like shape. The S_N2 reactions experience significant inhibition when the reagents have increased steric bulk, and they are only favored for primary substrates and straightforward secondary substrates. The S_N1 reaction involves the spontaneous dissociation of the substrate into carbocation, which then undergoes a second step where it reacts with a nucleophile. S_N1 reactions result in the racemization of configuration at the carbon atom and are primarily favored for tertiary substrates.

Elimination reactions frequently occur through three distinct mechanisms: E2, E1, and E1cB. These mechanisms vary in terms of the timing at which the breaking of C-X and C-H bonds takes place. The E2 reaction involves the simultaneous abstraction of a hydrogen atom by a base while the adjacent halide group departs. The E2 reaction typically yields a combination of alkene products, with the more highly substituted alkene being the dominant one, in accordance with Zaitsev's rule. In the E1 reaction, the initial step involves the cleavage of the C-X bond. The substrate undergoes spontaneous dissociation, resulting in the formation of carbocation. This carbocation then proceeds to lose H^+ from an adjacent carbon. In the E1cB reaction, the initial step involves the cleavage of the C-H bond. A base abstracts a hydrogen ion (H^+) to generate a carbanion intermediate, which is then followed by the departure of the halide group from the neighboring carbon atom. Biological elimination reactions commonly take place through the E1cB mechanism.

REFERENCES

Bruice, P. Y. (2015). *Essential Organic Chemistry* (Global Edition).

Clayden, J., Greeves, N., & Warren, S. (2012). *Organic Chemistry*. Oxford University Press.DOI: 10.1093/hesc/9780199270293.001.0001

Helmreich, R. F. (1953). *Mechanisms of Substitution and Elimination Reactions*. University of Colorado.

Kevill, D. N. (1959). *Substitution and Elimination Reactions in Non-hydroxylic Media*. University of London.

Klein, D. R. (2017). *Organic Chemistry*. John Wiley & Sons.http://books.google.ie/books?id=38aXDwAAQBAJ&pg=PA272&dq=substitution+and+elimination&hl=&cd=2&source=gbs_api

Liebman, J. F., & Greenberg, A. (1988). *Structure and Reactivity*. Wiley-VCH.

McMurry, J. (2003). *Fundamentals of Organic Chemistry*. Brooks Cole.

Mortimer, M., & Taylor, P. G. (2007). *Chemical Kinetics and Mechanism*. Royal Society of Chemistry.

Murray, P. R. S. (1977). *Principles of Organic Chemistry*. Heinemann.

P, N. R. (2011). *Fundamentals of Reaction Mechanisms in Organic Chemistry*. PHI Learning Pvt. Ltd.

Robertson, H. C. (2018). *On the Reactions of Alkyl Halides with Sodium Phenolate*. Palala Press.

Solomons, T. W. G., Fryhle, C. B., & Snyder, S. A. (2016). *Organic Chemistry*. John Wiley & Sons.

Solomons., Fryhle, C. B., & Snyder, S. A. (2022). *Organic Chemistry*. John Wiley & Sons.

Steigman, J. (1937). *The Walden Inversion and the Reactions of Alkyl Halides with Hydroxylic Solvents*. Columbia University.

Team, Open University an Introduction to the Chemistry of Carbon Compound (1973). *Substitution Reactions, a N D, Elimination Reactions, a N D, Addition Reactions*.

Walker, M. (2018). *Organic Chemistry 1*. State University of New York Oer Services.

Chapter 3
Proton–Coupled Electron Transfer in Organic Synthesis:
Mechanism and Application in Medicinal Chemistry

Zuber Peermohammed Shaikh
https://orcid.org/0009-0007-2606-8459
University of Pune, India

ABSTRACT

The coupled transfer of electrons and protons underlies a broad range of energy conversion processes. Proton-coupled electron transfer (PCET) reactions play an essential role in photosynthesis and respiration, as well as in many electrochemical reactions. Thus, understanding the fundamental theoretical principles of PCET reactions is critical for the design based on electrochemical reaction. A dipole-electric field coupling is quite reasonable as a model that captures the major effect of a polar solvent on the transferring proton. To investigate the analog of the solvent surfaces for electron transfer shown, we may use a four-state basis, where the base functions consist of electron and proton initial and final states (localized in the respective reactant and product wells for both the electron and the proton). This chapter elaborates the role of PCET reaction in human body like a proton pump and several other biogenic enzymes and general

DOI: 10.4018/979-8-3693-6473-4.ch003

(adiabatic) synthetic reactions of mechanism and applications of medicinal chemistry.

INTRODUCTION

The coupled transfer of electrons and protons is an important pathway of charge transport in a variety of biochemical, electrochemical, and small-molecule organic and inorganic reactions. The proteins and enzymes of photosynthesis and respiration have optimized structures that derive their function by utilizing energy gathered along a charge-separating network to drive a proton pump, which in turn is manifested in a transmembrane chemical potential that provides the energy for the synthesis of complex biomolecules. The quintessential electrochemical process of hydrogen-ion discharge at an electrode must involve a coupled electron and proton transfer. Recent developments in supramolecular solid-state chemistry are directed toward rational design of coupled electron proton transfer systems with novel electronic or photonic material properties. Here, photoinduced electron transfer induces proton transfer to promote a stable product. In this review, we outline our recent theoretical and experimental efforts aimed at a systematic study of the coupling between proton and electron transfer, which we refer to as proton-coupled electron transfer (PCET). The work has been directed toward understanding PCET in biomimetic systems, where the potential for this coupling can be assessed in a systematic manner. Naturally, these studies are also relevant to nonbiological PCET systems. (Graige MS, 1996)

The electron donor and acceptor are separated by a hydrogen-bonded interface. Photochemical initiation of the electron transfer reaction between the donor (e.g. Ru-bipy) and the acceptor (e.g. dinitrobenzene) may result in the transfer of a proton that, as indicated, results in a large charge change in the interface. These studies permit important PCET questions to be explored, such as the following: (a) What factors distinguish the consecutive process of electron transfer (ET) followed by proton transfer (PT) from a concerted transfer of an electron and a proton (ETPT)? (b) What structural/electronic features of the proton interface are important in governing the coupling between the electron and the proton? (c) How will the energetics (e.g. reorganization and free energy) for charge transfer in an ET or PT reaction be different in PCET? (d) How will the PCET rate compare in magnitude with ET and PT rates? We first discuss several examples of biological PCET that motivate the study of PCET in biomimetic compounds. PCET studies can be divided into those that address "physical" protons (those involved in translocation) and "chemical" protons (those needed for redox chemistry). Our focus is on the chemical protons that are tightly bound via hydrogen bonds to moieties that interact with electron transfer donors and acceptors (although, typically, later stages will involve the physical protons). An important PCET example is the photosystem II oxygen evolving complex (PSII/OEC), whose function is to split water. In bacterial reaction centers, the light-induced two-electron reduction of a quinone, QB, is coupled to proton transfer that ultimately leads to release of protons across the membrane to provide a transmembrane chemical potential. (Bögerhausen O, 1996)

Extensive efforts have been under way to resolve this overall two-electron process into its elementary steps of proton and electron transfer. Two quinones are involved: a tightly bound, primary quinone designated QA, and a loosely bound, secondary quinone, QB. The reaction of the second electron is tightly coupled to the proton, schematically (Q– AQ–B) + H+ → QA(QBH–), and can be formulated as ET followed by PT or vice versa, or as a concerted transfer. An attempt has been made to infer the reaction pathway by inserting QA's of differing driving force into the reaction center, and it was concluded that PT occurs first and is followed by a rate-limiting ET step. PCET also plays a crucial role at the terminus responsible for oxygen evolution. The OEC is linked to P680+ by a redox-active tyrosine, YZ, which functions as a charge transfer interface, allowing productive forward electron transfer reactions to compete successfully with the wasteful but thermodynamically favorable charge recombination reaction in the RC. Results by Meyer et al reveal that the protonation state of an acid-base group influences the kinetics of the YZP680+ → Y+ Z P680 reaction, which are consonant with Lavergne & Junge's results showing that proton release is coupled to the oxidation of YZ. EPR and ENDOR have verified that the oxidation of the tyrosine interface is closely coupled to proton transfer, but the mechanistic details of this PCET process remain elusive for YZ. Another tyrosine, YD, symmetrically disposed relative to

YZ, has been shown to be hydrogen bonded to a nearby histidine residue. The tryrosine is oxidized to a tryrosyl radical and re-reduces P680+ to P680. The proton in the tyrosylhistidine hydrogen bond transfers from the phenol to the nitrogen of the base. This combination of ET to form the tyrosyl radical and PT from the phenol to the histidine comprises a tightly coupled PCET system amenable to the type of analysis carried out below. In addition, PCET is not limited to charge transport at this terminus of the photosystem; it has been proposed that the oxidation of water at the oxygen-evolving complex (OEC) proceeds by a series of PCET reactions of an oxo-bridged cluster of manganese. (Lavergne J, 1993)

Figure 1. Mechanism of Proton Coupled Electron Transfer (PCET) reaction

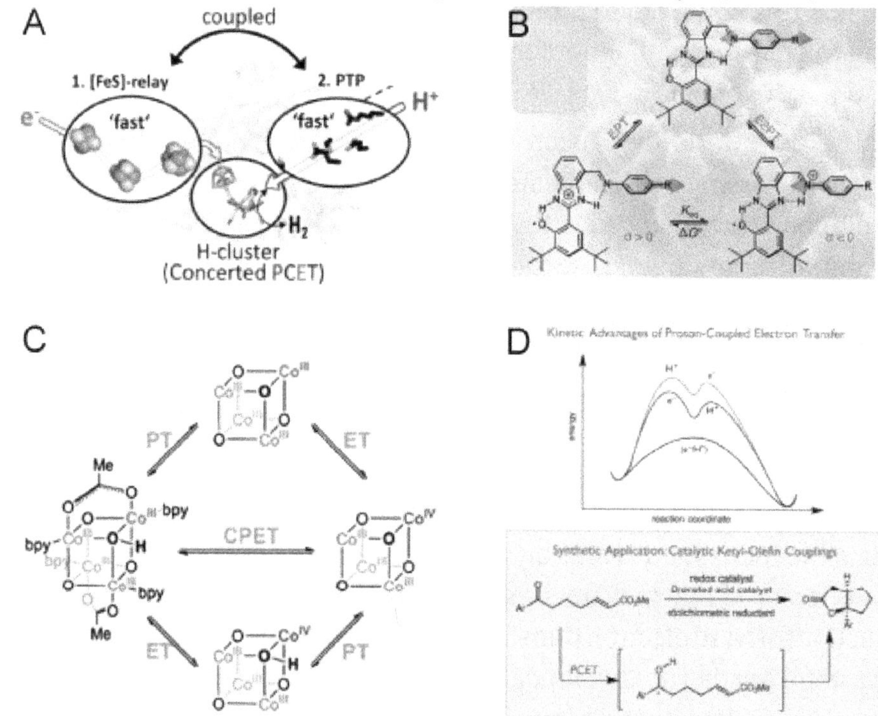

Consistent with this proposition is the marked pH dependence of the redox potential of OEC model oxo-manganese clusters. Careful and complete kinetic isotope effect studies by Meyer and co-workers on ruthenium-oxo dimers led to the conclusion that such reaction mechanisms involve simultaneous transfer of an electron and proton. We mention in passing the extensive studies of PCET in cytochrome c oxidase, where, for example, the rate of electron transfer from the heme a2+ to heme a3+ 3 is evidently proton coupled via interaction with a protonatable group close to the binuclear center. The role of proton influence on electron transfer in cytochrome c oxidase has recently been reviewed. Complexes between the nucleic acids in DNA and aromatic fluorophores exhibit strong fluorescence quenching of the latter by photoinduced electron transfer. (Brudvig GW, 1992)

The electron transfer quenching in aqueous medium is proton coupled, as inferred from isotope effects and the lack of quenching in nonaqueous solvent. The proton comes from the hydrogen bonding solvent, but the microscopic steps are difficult to elucidate, because the solvent contributes the proton. Molecular triads consisting of carotenoid–porphyrin– quinone (C−P−Q) triads have been designed with the goal of high quantum yield photochemically induced charge separation to Cq +−P−Qq. A recent synthetic design incorporates an intramolecular hydrogen bond from a carboxylic acid group to the quinone carbonyl. ET from the excited porphyrin results in a large increase in the pKa of the quinone. The consequent PT to the quinone carbonyl reduces the recombination rate and leads to a doubling of the quantum yield relative to the cognate nonhydrogen-bonded triads. Thus, for these triads, PT has a demonstrable effect on an ET reaction rate. PCET is not confined to the biological arena, of course. Hydrogen ion discharge on a metal surface, schematized as H3O+ + e−(Metal) = Hads + H2O, entails electron transfer from the metal with concomitant proton motion toward the electrode surface. (Roecker L, 1987)

LITERATURE REVIEW OF CURRENT CONTEXT OF RESEARCH

A recent theoretical article analyzed this reaction with some concepts similar to those we introduce in the next section and is based on the idea that the electron and proton transfer together. Two complicating features of this electrode reaction are the requirement of averaging the rate constant over the Fermi level of the electron in the metal (as in conventional electron transfer reaction treatments of electrochemical processes) and the difficulties engendered by the H3O+ not being a fixed distance from the electrode. An emergent theme in solid state materials chemistry is the synthesis of polymeric supramolecular structures with the goal of creating switchable materials. Recently, donor–acceptor dimers, capable of undergoing photoinitiated electron transfer, were shown to form a polymeric supramolecular structure held together by hydrogen bonds between the donor of one and the acceptor of another of the dimeric species. Upon optical excitation, charge transfer to form a radical ion-pair state with subsequent proton transfer in the hydrogen bond was postulated where the proton transfer step is required to stabilize the ion pair. Thus, electron transfer leads to proton transfer in these materials. Two-dimensional hydrogen bonded networks of quinhydrones have been synthesized. These H-bonded charge transfer complexes undergo electron and proton transfer, as inferred from the electronic spectrum of the product, but without kinetic studies the mechanism cannot be assessed. The desire to explore PCET in a well-defined geometry motivated the synthesis of model compounds that preserve the features of charge separating networks in biological systems and permit the interrogation of the electron and proton dynamics. (Binstead RA, 1992)

Figure 2. Resonating structures of Lewis acids of PCET

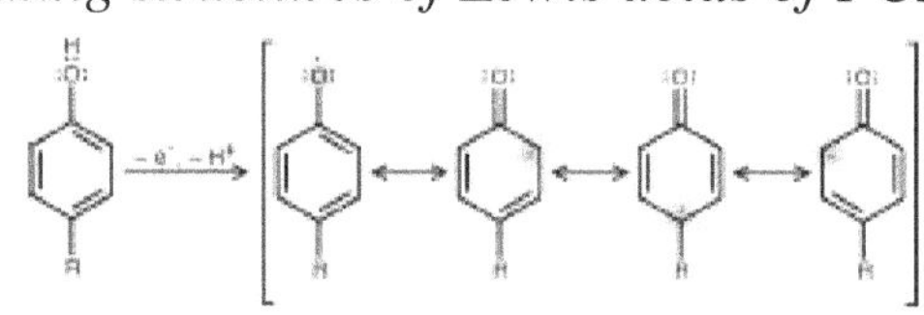

The key to the approach is to photoinduce electron transfer within a fixed-distance donor/acceptor pair that has a proton transfer network internal to the electron transfer pathway. The electron transfer kinetics is defined by color changes associated with the donor/acceptor chromophores, as monitored by time-resolved picosecond laser techniques. This approach is analogous to the progression in studies of ET reaction dynamics, where the original bimolecular reaction kinetic studies were difficult to interpret because of reaction at varying donor-acceptor distances, the necessity of evaluation of the energetics of the precursor and successor complexes, and interference from diffusion control for fast ET. This motivated the design of intramolecular ET systems, where the donor-acceptor distance is fixed and controllable, the energetics are specific to the ET step, and diffusion is not an issue. With the advent of photochemical initiation of ET, fast ET events could then be interrogated without the risk of a switchover to diffusion control of the reaction kinetics. This design for PCET follows a similar strategy, with a similar rationale. From the perspective of theory, fixing the geometry of the ET and PT donor and acceptors greatly simplifies the analysis. Averages over varying donor-acceptor distances do not have to be carried out, and since the reaction is intramolecular, the energetics are determined by the donor-acceptor hydrogen bonded complex's interaction with the solvent. The number of parameters that need to be evaluated once a rate expression is obtained is then greatly restricted relative to what would be required for a treatment of a bimolecular process. Then, the available theories for unimolecular ET and PT can be used as a framework with which to construct a PCET theory for consecutive reaction steps, and the construction of a theory for concerted ETPT does not suffer from such complications. (Kitagawa T, 1997)

Theory of PCET Reactions

In this section, we provide a theory for PCET in donor-acceptor complexes as exemplified. Our initial efforts were devoted to a description of PCET in symmetric interfaces, as experimentally realized by dicarboxylic acid hydrogen bonded interfaces. The asymmetric interfaces are of greater interest, as they are much more common in nature, and furthermore, the theory of asymmetric interfaces can of course be reduced to that of a symmetric interface by specialization. Thus, in this section, we restrict the discussion to asymmetric interfaces. The symmetric interface theory is summarized below, where the experimental results on carboxylic acid dimer interfaces are discussed. A key issue is the mechanism of the net reaction that transfers an electron and a proton from an initial state to a final state where both the electron and proton have transferred. In other words, the electron transfer's "function" is to transfer a proton. There are two, competitive, channels leading to both the electron and proton transferring. One channel is a consecutive process, with electron transfer (ET) followed by proton transfer (PT). This process is referred to as ET/PT, the slash denoting the consecutiveness of the process. (We assume that ET occurs first, as the experimental design discussed here initiates ET photochemically. In any case, the process of PT followed by ET can be discussed, formally, in similar fashion to ET/PT.) Let us stress that the electron transfer may or may not stimulate the proton transfer. Upon electron transfer, the proton may be thermodynamically and/or kinetically prompted to transfer as the electronic energy and solvation energetics that the proton experiences will typically be different after the ET reaction. That is, the potential energy surface (pes) governing the proton transfer may become favorable for proton transfer after the electron transfer. Thermodynamically, the proton may exhibit a lower pKa after ET. But to predict kinetics, the driving force of the reaction, 1G(0), is not sufficient. For proton transfer reactions whose mechanism can be formulated through Marcus-Levich theory, there will be a reorganization energy λs contribution to the rate expression, and this is a nonequilibrium quantity. ET may stimulate PT is a feature analogous to excited state proton transfer reactions (photoacids) where, upon electronic excitation, a proton may become much more labile. Here, it is the electron transfer that may act like an electronic excitation in the sense of inducing a proton transfer reaction. In ET/PT there are two quantum events: electron tunneling and proton tunneling. (Gust D, 1991)

These tunneling events are both induced by solvent fluctuations. (Contributions from bond rearrangements can also occur upon the electron transfer, but in many reactions, solvation energetics is dominant). For electron transfer, the Marcus-Levich picture is schematized, from the point of view of the electron coordinate (re) and the solvent coordinate (R). The solvent fluctuates to a configuration where the electron's energy on the donor (D) and acceptor (A) are essentially equal. At this "transition state" configuration, the electron can tunnel through the potential barrier and then, via further solvent fluctuations, be stabilized in the product state. The solvent coordinate figure is obtained by expressing the Hamiltonian consisting of the terms for the electron, its interaction with the solvent, and the solvent's energy in a two-state basis of localized (initial and final) states. The diabats for the initial electron state, i, and the final electron state, f, are drawn. Including the off-diagonal coupling, responsible for the charge transfer, produces adiabatic surfaces that split the diabats in the region of their crossing. The separation between the two (symmetric and antisymmetric) eigenfunctions is twice the electronic coupling matrix element. (Gust D, 1993)

Figure 3. Surface free energy of PCET reactions

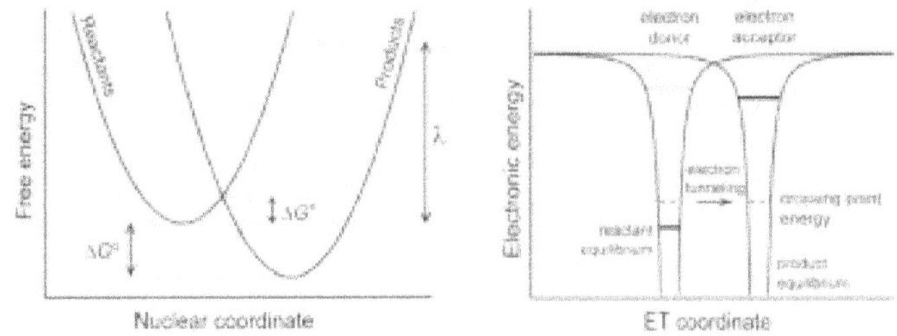

It represents the electronic overlap of the wavefunctions on donor and acceptor. In a simplest ET picture, then, the rate will be determined by this electronic overlap, and the activation energy defined as the fraction of all solvent configurations that correspond to this transition state. The three surfaces correspond to solvent configurations R1 (R∗) [R2] with the electron in its initial (transition) [final] state. With the solvent at R∗, the electron may transfer. (b) Marcus solvation (nuclear) coordinate for an ET reaction. rate constant is given by kET = V 2 el h¯2 q πh¯2 ± λskB T e−(λs+1G(0))2/4λs kB T . 1. The Marcus form of the activation energy above is obtained by a classical treatment of the solvent (its characteristic frequencies should be small compared with kBT), and the assumption that the solvation surfaces are quadratic. The latter assumption is a consequence of the assumed linear response of the solvent to the presence of the charge distribution of the solute. Proton transfer reactions can be thought of in similar fashion. However, it must be stressed that there are other mechanisms for a proton transfer reaction. We assume here that the proton transfer mechanism is the same as that for electron transfer. Weaker H bonds favor this circumstance, indicating a substantial barrier between the initial and final states. An expression analogous to Equation 1 may be used for kPT. We denote the proton's initial state as a final state, when it has transferred, as b. The consecutive mechanism is distinct from another pathway for the overall conversion: a concerted process whereby the electron and the proton transfer during the same tunneling event. In this case, a given solvent fluctuation produces a state with a significant probability for the electron and proton to transfer together. We refer to this process as ETPT. We have formulated an approach to ETPT that is analogous to that outlined above for ET or PT and refer to it as the two-dimensional (2D) approach. In this method, the electron and the proton are treated on the same footing in a 2D tunneling space, where the energy levels are those of the proton and electron, and these levels are parametric on the solvent configuration. It is a generalization of the ET mechanism sketched above. With this approach, we may obtain a qualitative picture of the ingredients that lead to ET/PT or ETPT. A suitable potential energy function for this discussion is

V(re,r p I R) = V(re) + V(r p) + γ rer p + cereR + cpr pR + 1 2 ks R2

The terms on the right side of Equation represent, respectively, the isolated transfer-electron's pes with re, the electron's coordinate; the isolated transfer-proton's pes with rp, the proton's coordinate; a coupling between the electron and proton with strength γ, the coupling between the solvent, with coordinate R, and the electron (proton), with strength ce(cp); and the solvent's contribution to the energetics with ks, the solvent force constant. In the absence of the electron proton coupling term (γ = 0), the Hamiltonian based on Equation would be separable and each part would lead, in the two-state basis approach discussed above, to the charge transfer mechanism schematized. (Krishtalik LI, 1970)

Figure 4. Effect of solvent and proton transfer in PCET reaction

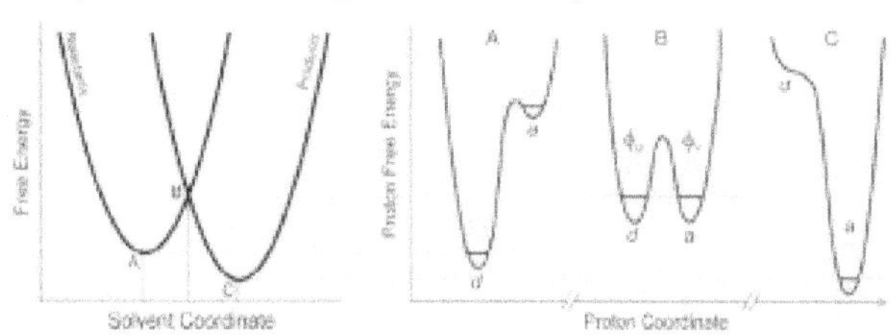

Dielectric Theory of PCET

The use of a harmonic approximation for the solvent is familiar from dielectric theory where the appropriate coordinate is the solvent's orientational polarization. With suitable interpretation of the force constant, it is Marcus's reaction coordinate. The transfer electrons and proton's potential can be thought of as symmetrical double-well potentials with minima at −reo (+reo) for the i (f) state and at −rpo (+rpo) for the proton a (b) state. The coupling γ between the electron and proton serves as an electronic structure effect whereby the electron transfer reaction may make proton transfer more favorable thermodynamically. For example, if γ were negative, electron transfer would be endothermic and the subsequent proton transfer exothermic, while the ETPT event would be thermoneutral. The functional form of the interaction is not of significance. Whether this coupling does enhance PT upon ET or enhances ETPT will also depend on the solvation energetics of the corresponding reaction. The terms proportional to ce (cp) provide the effect of the solvation coupling to the electron (proton) that can symmetrize the pes to permit electron, proton, or electron-proton transfer. In the contribution cprp R, rp may be regarded as the dipole moment operator of the proton that changes as the proton transits from its initial to final state; R would then be the fluctuating electric field from the solvent, and cp would determine the strength of the coupling. In this localized-state basis (not the eigenstates of H), the four energies that are parametrically dependent on the solvent coordinate R correspond are the following diabatic surfaces. (Benderskii VA, 1994)

BACKGROUND OF CURRENT CONTEXT OF RESEARCH

The coupled transfer of electrons and protons underlies a broad range of energy conversion processes. Proton-coupled electron transfer (PCET) reactions play an essential role in photosynthesis and respiration, as well as in many electrochemical reactions. Thus, understanding the fundamental theoretical principles of PCET reactions is critical for the design based on electrochemical reaction artificial photosynthesis. Recent advances in the theory of PCET enable the prediction of the impact of system properties on the reaction rates. These predictions may guide experimental efforts in tuning systems to enhance energy conversion processes. This Account summarizes the theoretically predicted dependence of PCET rates on system properties and illustrates potential approaches for tuning the reaction rates in chemical systems. Here PCET reactions refer to the transfer of an electron and a proton in a single step without a stable intermediate. Such reactions are often denoted concerted, although the term concerted is not rigorously applicable because the electron and proton behave quantum mechanically and hence are delocalized. (Shafir A., 2016)

The PCET rate depends on the reorganization energy, the driving force, the electronic coupling, and the proton transfer interface properties, such as the equilibrium proton donor-acceptor distance and frequency. In addition to these properties that are directly related to the redox species, the PCET rate also depends on temperature, pH, overpotential in electrochemical processes, and solvent relaxation time in certain regimes. Overview of PCET Theory Typically PCET reactions are in the vibronically nonadiabatic regime because the couplings between the diabatic reactant and product vibronic states are significantly smaller than the thermal energy. In this regime, PCET reactions are described in terms of nonadiabatic transitions between the reactant and product electron-proton vibronic states. The reactant electronic state corresponds to the electron localized on its donor, and the product electronic state corresponds to the electron localized on its acceptor. When the proton vibrational states are calculated for the proton potential energy curves associated with the reactant and product electronic states, the reactant and product electron-proton vibronic states are products of the corresponding electronic and proton vibrational states. In the limit of electronically non adiabatic proton transfer, the coupling between a pair of vibronic states is the product of an electronic coupling and the overlap integral between the reactant and product proton vibrational wave functions. In general, the overlap integral depends strongly on the proton donor-acceptor distance. As a result, the inclusion of the proton donor-acceptor vibrational motion is important for PCET reactions. The fundamental mechanism of a typical PCET reaction is illustrated, which depicts slices of the ground state free energy surfaces along a collective solvent coordinate, in conjunction with the proton potential energy curves and the corresponding proton vibrational wave functions. As shown in this figure, the proton vibrational ground state wave function is localized near the donor for the reactant state and near the acceptor for the product state. Changes in the collective solvent coordinate (i.e., reorganization of the solvent environment) strongly impact the relative energies of the reactant and product proton potential energy curves. For a typical equilibrium PCET reaction, the system is initially at thermal equilibrium on the reactant surface, and fluctuations of the solvent environment bring the system to the crossing point between the two surfaces, where the reactant and product vibronic states are degenerate. The probability of a non-adiabatic transition from the reactant to the product surface is proportional to the square of the vibronic coupling, which in turn is proportional to the proton vibrational wave function overlap depicted. Subsequent to such a nonadiabatic transition, the system relaxes to thermal equilibrium on the product surface. In general, the excited vibronic states may also play an important role in PCET reactions. (Pulis A.P., 2016)

METHODOLOGY OF CURRENT CONTEXT OF RESEARCH

We have derived a series of non-adiabatic rate constant expressions for PCET reactions in various well-defined limits. These expressions are based on Fermi's golden rule formalism, which is valid in the vibronically nonadiabatic limit, in conjunction with linear response theory. The simplest rate constant expression is for fixed proton donor-acceptor distance R and thus is valid for stiff hydrogen bonding interfaces. The more complete rate constant expression includes the motion of the R-mode, which is represented by a quantum mechanical harmonic oscillator associated with the reduced mass M and frequency Ω. We derived simplified versions of this rate constant expression in the low- and high frequency regimes for the R-mode. The low-frequency (high temperature) R-mode rate constant expression is based on the assumption that pΩ, kBT, and the high-frequency (low-temperature) R-mode rate constant expression is based on the assumption that pΩ . kBT. The rate constant expression for fixed proton donor-acceptor distance R is10 where the summations are over reactant and product vibronic states, Pμ is the Boltzmann probability for the reactant state μ, and Vel is the electronic coupling. In addition, Sμν (0) is the proton vibrational wave function overlap, λμν is the total reorganization energy, and ΔGμν 0 is the free energy of reaction for vibronic states μ and ν. Often the reorganization energy λμν is assumed to be the same for all pairs of states, and ΔGμν 0 is assumed to be the sum of the driving force ΔG0 ≡ ΔG00 0 and the difference between the product and reactant proton vibrational states relative to their respective ground states. When the proton donor-acceptor motion is included, the overlap is approximated as Sμν) Sμν (0) exp[-Rμν(R - R¯)], where R¯ is the equilibrium value of R and Sμν (0) is the overlap at R) R¯. The rate constant expression in the low-frequency (high-temperature) R-mode regime is similar to eq 1 but includes the extra prefactor exp[2kBTRμν2/MΩ2], and the total reorganization energy includes two additional terms due to the dependence of the vibronic coupling on R and the reorganization energy corresponding to the R-mode. The rate constant expression in the high-frequency (low-temperature) R-mode regime is similar but includes an extra prefactor that depends on the frequency Ω but not on the temperature. The form of the rate constant is valid for vibronically nonadiabatic reactions, where the solvent time scale is much faster than the time scale associated with nonadiabatic transitions between vibronic states. (Yanagi T., 2018)

The specific form of the vibronic coupling is valid for electronically nonadiabatic proton transfer, where the proton tunneling time scale is significantly faster than the electronic transition time scale. In the opposite regime of electronically adiabatic proton transfer (e.g., traditional hydrogen atom transfer), the vibronic coupling is no longer the product of the electronic coupling and the proton vibrational wave function overlaps. The kinetic isotope effect (KIE), which is the ratio of the rate for hydrogen to the rate for deuterium, provides a useful mechanistic probe for PCET reactions. Based on the rate constant given in eq 1 and related expressions, the KIE is proportional to the ratio of the overlaps of the hydrogen and deuterium vibrational wave functions for a given pair of reactant/product vibronic states. Typically, this ratio of overlaps increases as the overlaps decrease. Thus, this ratio is larger for greater proton donor-acceptor distances and is smaller for excited vibronic states than for the ground vibronic states because the excited state proton vibrational wave functions are more delocalized. The magnitude of the KIE has been observed experimentally to range from moderate to extremely large (e.g., >50). As will be discussed below, analysis of the magnitude and temperature dependence of the KIE provides insight into the underlying PCET mechanism. We have also extended this theoretical formulation to electrochemical PCET. Specifically, we have derived expressions for the heterogeneous rate constants and current densities for processes in which proton transfer occurs within a solvated hydrogen-bonded solute complex concurrently with electron transfer between the solute complex and the immersed electrode. The resulting expressions for the transition probabilities are similar to those derived for homogeneous PCET with the reaction free energy replaced by an expression that depends on the energy of the one-electron acceptor state in the electrode, the potential at the redox species in solution due to the electrical double layer formed by the supporting electrolyte ions, and the overpotential. The heterogeneous rate constant expressions are obtained by integration over the energy levels in the conduction band of the electrode, weighting the transition probability according to the Fermi distribution and the density of states of the electrode. The anodic and cathodic current densities are obtained by integrating the product of the rate constant and the relevant solute complex concentration over the distance between the solute complex and the electrode surface. Recently we have extended the theory of PCET beyond the golden rule formalism and derived rate constant expressions that include the effects of solvent dynamics in both homogeneous and electrochemical PCET. These expressions interpolate between the golden rule and solvent-controlled limits, where the golden rule limit is defined in terms of weak vibronic coupling and fast solvent relaxation, and the solventcontrolled limit is defined in terms of strong vibronic coupling and slow solvent relaxation. In both cases, the overall k) ∑ μ Pμ∑ν |VelSμν (0)| 2 p π λμνkBT exp[- (ΔGμν 0 + λμν) 2 4λμνkBT] (1) Proton-Coupled Electron Transfer Hammes-Schiffer treatment is nonadiabatic in the sense that transitions between electronic states are involved. Consistent with our previous work, the rate constant is proportional to the square of the vibronic coupling and is independent of the solvent relaxation time in the golden rule limit. (Kaiser D., 2019)

One drawback of the reaction of unprotected anilines is its limited scope. To achieve a more general method, we investigated the effect of protecting groups on the aniline substrates. Electron- withdrawing protecting groups on the nitrogen should impart several favorable features, for example, reduce the nucleophilicity of the nitrogen to avoid direct reaction with anhydride, increase the acidity of the N–H unit to construct an S–N bond by concerted deprotonation (from INT1 to INT2), and weaken the S–N bond in the precursor to charge accelerated *sigmatropic rearrangement.* After the screening of electron-withdrawing groups, sulfonyl groups were found to participate in the biaryl formation efficiently (similar to the indole synthesis), and p-methoxybenzenesulfonyl group showed the best performance among those tested. While the scope of the reactions is not so wide as in the case of phenols, a range of heteroaryl sulfoxides and naphthyl sulfoxides reacted to yield sulfonylaminobiaryls, (He, Z., 2019)

Synthesis of 2-Hydroxy-2′-Iodobinaphthyls

We also sought sulfur-free surrogates of aryl sulfoxides for biaryl synthesis. Considering the similarity in structure and reactivity we focused on hypervalent iodine compounds. However, because trivalent iodine compounds are strong oxidants, we naturally suffer from oxidative side reactions of phenols. After extensive screening, we found that 2-naphthols are suitable for the coupling and that treatment with 2-iodonaphthalene diacetate in acetic acid yielded 2-hydroxy-2B-iodobinaphthyls. Furthermore, iodoarene diacetate can be generated in situ and used without isolation; iodobenzene 45 was treated sequentially with mCPBA and 2-naphthol to afford biaryl.

RESULT OF CURRENT CONTEXT OF RESEARCH

In contrast, the rate constant is independent of the vibronic coupling and increases as the solvent relaxation time decreases in the solvent-controlled limit. The interconversion between these two limits can be induced by altering the proton transfer interface properties (i.e., the proton donor-acceptor mode frequency and the proton vibrational wave function overlap), the electronic coupling, the solvent relaxation time, and the overpotential in the case of electrochemical PCET. Moreover, the KIE behaves differently in the golden rule and solvent-controlled limits and thus provides a unique probe for distinguishing between these two different regimes. As discussed above, the KIE can be quite large in the golden rule limit because it is approximately proportional to the square of the ratio of the overlaps of the hydrogen and deuterium vibrational wave functions. The KIE is expected to be more moderate in the solvent-controlled limit because it arises mainly from differences in zero-point energy and vibronic energy level splitting. Dependence of PCET Rates on System Properties Analysis of the rate constant expressions described above provides insight into the underlying physical principles of PCET and enables the prediction of the dependence of the rates on physical properties of the system. In general, the PCET rate will increase as the electronic coupling increases and as the total reorganization of energy decreases. In the solvent-controlled limit defined above, the PCET rate will increase as the solvent relaxation time decreases. The dependence of the PCET rate on the proton transfer interface properties, driving force, temperature, pH, and electrochemical potential is more complicated and will be discussed in the remainder of this section. The rates of PCET reactions are strongly influenced by the properties of the proton transfer interface. The rate increases dramatically as the equilibrium proton donor-acceptor distance decreases due to the increase in the proton vibrational wave function overlap. Furthermore, typically the rate increases as the proton donor-acceptor mode frequency decreases because the lower frequency enables sampling of smaller distances corresponding to larger overlaps. This effect is evident in the prefactor exp[2kBTRμν2/MΩ2] that arises in the low-frequency R-mode regime. (Zhang L., 2019)

Figure 5. Electronic configuration of hetrocyclic adiabatic reaction in PCET reaction

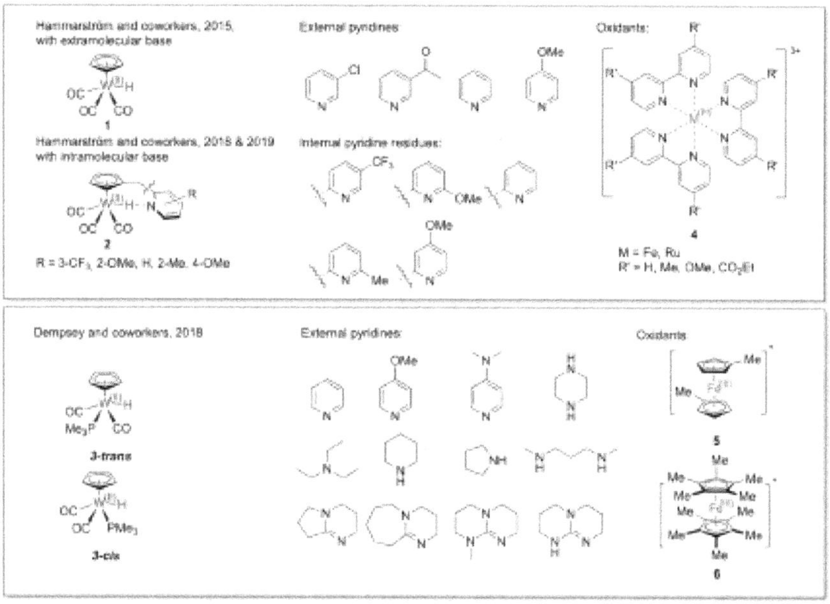

Often the ***proton donor-acceptor mode*** frequency decreases as the equilibrium distance increases, so these two trends may work against each other when designing proton transfer interfaces aimed at enhancing the PCET rate. The KIE is also strongly influenced by the proton transfer interface properties. As mentioned in the previous section, the KIE is approximately proportional to the square of the ratio of the hydrogen and deuterium vibrational wave-function overlaps. For a given pair of reactant/product vibronic states, the KIE increases as the equilibrium proton donor-acceptor distance increases when all other parameters remain the same. Moreover, in the low-frequency R-mode regime, the KIE increases as the proton donor-acceptor mode frequency increases due to the prefactor mentioned above. Again, these two trends may work against each other because the proton donor-acceptor mode frequency often decreases as the equilibrium distance increases. In addition, the rates of PCET reactions depend strongly on the driving force. On the basis of eq 1 with only the ground vibronic states included, the rate constant would exhibit a turnover as ΔG0 becomes more negative, with a maximum at -ΔG0) λ, as depicted. This turnover is similar to the inverted region's turnover predicted by Marcus theory for electron transfer. As indicated by the solid line, however, the inclusion of excited vibronic states leads to qualitatively different behavior. In particular, the rate constant will increase as $\Delta G0$ becomes more negative because excited vibronic product states associated with low free energy barriers and relatively large vibronic couplings become accessible. The PCET rate constant will eventually decrease for highly exoergic reactions that are expected to be experimentally inaccessible. Note that ET reactions coupled with low-frequency vibrational modes can exhibit an inverted regional turnover at experimentally accessible driving forces. This behavior is not predicted for PCET reactions mainly because of the high frequency of the proton vibrational motion, as well as the greater shift of the proton vibrational wave function. (Higuchi K, 2021)

Moreover, the excited vibronic states also impact the qualitative behavior of the driving force dependence of the KIE. The ln(KIE) exhibits a very weak quadratic dependence on driving force, with a maximum at $\Delta G0 \approx 0$, when only the ground states are included. In contrast, the ln(KIE) decreases much more rapidly for both positive and negative driving forces when excited vibronic states are included because the contributions from excited vibronic states increase as the reaction becomes more asymmetric, and contributions from excited vibronic states decrease the KIE. These illustrate that the (0/0) pair dominates at $\Delta G0 \approx 0$, whereas other pairs dominate sequentially as the reaction becomes more endergonic or exergonic. Similar to most chemical reactions, PCET rates increase with temperature. The temperature dependence of the rate is dictated mainly by the exponential term in eq 1, which typically overrides the temperature-dependent prefactor. Because of contributions from excited vibronic states, the increase of the rate with temperature may not behave as a single exponential. Moreover, in the low-frequency R-mode regime, the additional temperature-dependent exponential term also increases with temperature. The KIE, however, can exhibit more interesting temperature dependence. The temperature dependence of the KIE has been measured experimentally to probe the detailed mechanism of PCET reactions. (Yoshida, S., 2009)

Assuming that the driving force and reorganization energy are independent of isotope and only the ground states contribute to the overall rate, the rate constant expression in eq 1 for fixed R predicts that the KIE will be independent of temperature. Using the same assumptions in the low-frequency R-mode regime, the KIE is predicted to decrease with temperature as $\exp[-2kBT(RD\ 2 - RH\ 2)/M\Omega 2]$, where $RD > RH$ because the overlap decreases faster with R for deuterium than for hydrogen. This factor tends to dominate the temperature dependence of the KIE in the low-frequency R-mode regime even when excited vibronic states are involved. Since this temperature-dependent factor is not present in the high-frequency R-mode regime, however, the KIE could either increase or decrease with temperature in this regime when excited vibronic states contribute significantly to the overall rate. A number of experimental studies have illustrated that the rate of PCET reactions depends strongly on pH. The basis for this pH dependence has been discussed extensively in the literature. When the proton acceptor is a buffer species in solution, the pH dependence is thought to arise from the protonation equilibrium of the buffer. The pH dependence of tyrosine oxidation for one specific ruthenium-polypyridyl-tyrosine system has been observed to be independent of the buffer, however, and has been interpreted in terms of proton transfer to bulk water. The physical basis for the pH dependence observed in this system is not fully resolved. In general, the pH dependence of PCET reactions could arise from the kinetic complexity of competing concerted and sequential PCET reaction pathways, as well as pathways involving different proton acceptors, including various buffer species and hydroxide. Moreover, the pH dependence could be caused by the impact of pH on the physical properties of the proton transfer interface, such as hydrogen bond strength, proton donor-acceptor distance, and frequencies. Experimental verification of these hypotheses is challenging. In electrochemical PCET, the heterogeneous rate constants and current densities depend strongly on the electrode potential. Similar to homogeneous PCET, the current densities also depend on the proton transfer interface properties, reorganization energy, temperature, and pH. In addition, the effective activation energies associated with the current densities. (Huang X., 2012)

The difference in rates of proton transfer along the "fast" pumping channel leading from Glu242 to His291 and the "slow" chemical channel leading from Glu242 to the BNC presumably is based on the structural arrangements of the key groups. The protonated Arg438 is located in the immediate vicinity of His291, so that when the proton is needed on

His291, Arg438H+ quickly donates the proton to this residue. This proton transfer is achieved via a water molecule Wa3. Immediately after that, the chain of water molecules connecting the PRDa3 and Arg438 site with Glu242 provides a proton from Glu242 to reprotonate Arg438. The net result of these transitions is that a proton. (A) Key structural elements of the proposed pumping mechanism of CcO and the sequence of transitions during one pumping cycle. Two protonation sites (the PLS and a site in the BNC) are shown as H-circles. PT and ET steps are shown by blue and red arrows, respectively. The residue notation is for bovine enzyme. The key assumption of the model is that, upon ET between the hemes (step 2), the proton transfer to the proton loading site (step 3) occurs before the proton transfer to BNC (step 5). Hammes-Schiffer and Stuchebrukhov from Glu242 is quickly transferred to His291. The exact mechanism of gating, however, is still unknown. Recently, Wikstro¨m and co-workers proposed that water dynamics in the catalytic cavity might in fact be responsible for proton gating. For proper function of the pump (i.e., to pump protons in the right direction, against the membrane electrochemical gradient), another key requirement should be fulfilled, namely, that the protonation of the PLS occurs by a proton from the negative side of the membrane with low chemical potential and not from the opposite side with high chemical potential. Otherwise, the protons would flow in the wrong direction. Closely related to this requirement is the one that is usually assumed in the form of a mechanical gate that would prevent the leak of protons through the pump between the pumping events. The two-step pumping mechanism, in which the pumped proton is first loaded to the PLS from the N-side of the membrane and then expelled by the chemical proton arriving to the BNC on the P-side of the membrane, is part of many recently discussed models. The main difference between the models is the nature of the PLS site, whose identity is still not known; however, recent experimental and theoretical studies point to one of the residues in the group His291, propionate A or D of heme a3, or a group (including a water molecule) nearby. TfOH yielded 2-amino-2B-sulfanylbiaryl in good yield. (Eberhart A.J., 2011)

DISCUSSION OF CURRENT CONTEXT OF RESEARCH

HAT vs PCET Mechanisms

In general, HAT reactions can be viewed as a subset of the concerted PCET reactions described. Nevertheless, to aid in discussions of chemical reactions, it is often useful to distinguish between these two types of mechanisms. A textbook example of the fundamental differences between the PCET and HAT mechanisms is provided by the comparison between the phenoxyl/phenol and benzyl/toluene self-exchange reactions. These systems have been analyzed in terms of the singly occupied molecular orbitals (SOMOs) obtained from density functional theory calculations of the transition state structures (i.e., the first order saddle points). As depicted, the SOMO is dominated by 2p orbitals perpendicular to the proton donor-acceptor axis for the phenoxyl/phenol system, but the SOMO is dominated by atomic orbitals oriented along the proton donor-acceptor axis for the benzyl/toluene system. These results were interpreted to signify that the electron and proton are transferred between different sets of orbitals for the former system but between the same sets of orbitals for the latter system. This interpretation led to the identification of the phenoxyl/phenol reaction as PCET and the benzyl/toluene reaction as HAT. These systems have also been analyzed with the semiclassical formalism described. The electronically diabatic potential energy curves corresponding to the reactant and product PCET states defined were obtained by fitting to the CASSCF (complete active space self-consistent-field) electronically adiabatic ground- and excited state potential energy curves. Mixing the two diabatic states with the appropriate electronic coupling leads to the CASSCF electronically adiabatic curves, as depicted. The parameters comprising the electronic transition time τe and the proton tunneling time τp defined, as well as the proton adiabaticity parameter p) τp/τe, were obtained from these diabatic potential energy curves. As discussed, the degree of electronic nonadiabaticity for the proton transfer reaction provides a quantitative diagnostic for distinguishing between HAT and PCET. Specifically, the PCET mechanism corresponds to the electronically nonadiabatic limit (p,1), and the HAT mechanism corresponds to the electronically adiabatic limit (p.1). (Eberhart A.J., 2013)

These calculations revealed that the phenoxyl/phenol reaction involves electronically nonadiabatic proton transfer and corresponds to PCET, whereas the benzyl/toluene reaction involves electronically adiabatic proton transfer and corresponds to HAT. For the phenoxyl/phenol system, the adiabaticity parameter p) 0.013, which is in the electronically nonadiabatic limit with $\tau e \approx 80\tau p$. In this case, the electronic transition time is significantly greater than the proton tunneling time, so the electrons are not able to rearrange fast enough for the proton to remain on the electronically adiabatic ground state surface. For the benzyl/ toluene system, the adiabaticity parameter p) 3.45, which is found to be in the electronically adiabatic limit with $\tau e \approx 0.29\tau p$. In this case, the electronic transition time is less than the proton tunneling time, so the electrons can respond instantaneously to the proton motion, allowing the proton to remain on the electronically adiabatic ground state surface. This characterization of the phenoxyl/phenol system as PCET and the benzyl/toluene system as HAT is consistent with the qualitative analysis based on the SOMOs. Further analysis of the semiclassical calculations provides insight into the fundamental differences between these two systems. The electronic coupling Vel was estimated to be half the splitting between the two electronically adiabatic CASSCF potential energy curves at the midpoint between the proton donor and acceptor atoms. From this electronic coupling was found to be 700 cm-1 for the phenoxyl/phenol system and 14,300 cm-1 for the benzyl/ toluene system. Note also that the diabatic states exhibit greater slopes at the crossing point for the former system than for the latter system, resulting in a larger $|\Delta F|$ for the phenoxyl/phenol system. The smaller Vel and larger $|\Delta F|$ lead to a smaller value of the proton adiabaticity parameter for the phenoxyl/phenol system. As a result of these differences, the vibronic coupling has different forms for these two systems: it is the product of the electronic coupling and the overlap of the reactant and product proton vibrational wave functions for the phenoxyl/ phenol system, but it is half the tunneling splitting on the electronically adiabatic ground state for the benzyl/toluene system. We point out that the splitting between the ground and excited electronic states are significantly larger than the thermal energy kBT at room temperature for both systems. The significance of electron-proton nonadiabatic effects is not determined by comparing this splitting to the thermal energy but rather by comparing the electronic transition time to the proton tunneling time. In contrast, the overall vibronic couplings for both systems were found to be much smaller than the thermal energy, mainly due to contributions from the proton vibrational wave functions, indicating that both reactions are vibronically nonadiabatic with respect to a solvent or protein environment. Thus, the types of rate constant expressions are applicable to these systems. (Eberhart A.J., 2013)

Concerted PCET in Solution and Enzymes

The parameters in the rate constant expressions given in section 2 can be determined using various experimental and theoretical methods. The driving force can be estimated from experimentally measured redox potentials and pKa values. The solvent reorganization energy of a PCET reaction is often similar to that of the corresponding ET reaction because the solvent reorganization energy for PT is typically much smaller than that for ET (i.e., the proton transfers a much shorter distance than the electron). In principle, the equilibrium proton donor-acceptor distance could be obtained with experimental methods for structure determination, and the associated frequency could be determined spectroscopically, although the separation of this mode from other modes is challenging. In practice, the equilibrium proton donor-acceptor distance and frequency, as well as the proton potential energy curves, are usually obtained from electronic structure calculations. Alternatively, molecular dynamics simulations may be used to estimate this distance and frequency. The reactant and product proton vibrational wave functions and the associated overlaps can be calculated for the proton potential energy curves using Fourier grid methods. The electronic coupling can be calculated with the same electronic structure methods that are used for calculating this parameter in electron transfer theory. Note that the magnitude of the electronic coupling is usually similar for related PCET and ET reactions, but the PCET electron-proton vibronic coupling is substantially smaller due to inclusion of the overlap between the reactant and product proton vibrational wave functions. (Kobatake, T., 2010)

Currently, methods for calculating the combined electron-proton vibronic coupling in terms of mixed nuclear-electronic wave functions are being developed. The formulation described has been used to predict the dependence of the concerted PCET rate constants and KIEs on system properties, including temperature and driving force, as well asthe equilibrium proton donor-acceptor distance and frequency. From the experimental standpoint, altering only a single parameter without influencing the other parameters is often not possible. For example, an increase in the equilibrium proton donor-acceptor distance is often associated with a decrease in the frequency. In addition, changing the driving force for proton transfer often impacts the equilibrium proton donor-acceptor distance. The driving force dependence of the rate constant is particularly interesting in the context of the Marcus inverted region behavior, where the rate constant decreases as the driving force increases (i.e., as the reaction becomes more exoergic). Inverted region behavior has been observed experimentally for electron transfer reactions. Inverted region behavior is predicted to be experimentally inaccessible for concerted PCET reactions because of the availability of excited electron-proton vibronic product states with greater vibronic coupling. As illustrated for model systems, however, apparent inverted region behavior could be observed if varying the driving force also impacts other properties of the system, such as the proton donor-acceptor distance. (Huang X., 2011)

The theory described has been applied to a variety of experimentally studied concerted PCET reactions in solution and enzymes. These applications include PCET in amidinium-carboxylate salt bridges, iron bi-imidazoline complexes, ruthenium-polypyridyl complexes, ruthenium-polypyridyl-tyrosine systems, rhenium-polypyridyl-tyrosine systems, thymine-acrylamide complexes, and ruthenium-polypyridyl-quinol systems. This theory has also been applied to PCET in the enzyme soybean lipoxygenase, as well as to a series of mutant forms of this enzyme. Furthermore, the extension of this theory to electrochemical systems has been applied to osmium complexes attached to a self assembled monolayer on a gold electrode. Three of these systems are depicted. All of these calculations have reproduced the experimentally observed trends in the rates and KIEs, as well as the temperature and pH dependences in some cases. In addition, these studies have elucidated the fundamental principles underlying PCET reactions, assisted in the interpretation of experimental data, and provided experimentally testable predictions. 6.3. Cytochrome c Oxidase A marvelous illustration of many aspects of coupled electron and proton transfer reactions in biology is provided by cytochrome c oxidase. This enzyme is a redox-driven proton pump that utilizes the energy of oxygen reduction to pump protons across the membrane. The A-type oxidases have two proton-conducting input channels (D- and K-channels) to transfer protons required for oxygen reduction and for proton pumping, with the D-channel transporting all pumped protons. Electrons are transferred along a chain of metal cofactors: from cyt c to CuA, then to heme a, and finally to the Fea3-CuB binuclear catalytic center of the enzyme. The electron transport is coupled to proton translocation in such an intricate way that as electrons flow to the catalytic center of the enzyme to accomplish oxygen reduction (and produce water), the protons are pumped across the membrane against the electrochemical proton gradient. (Huang X., 2013)

Proton Pumping Mechanism

Since the discovery of the proton pumping by CcO in 1977 by Wikström, there were many proposals for the mechanism of the enzyme. Some computational studies point to the mechanism of pumping shown. Recent experiments and their modeling support the mechanism in its general form; however, some details (i.e., the identity of the so-called proton loading site; see below) remain experimentally untested and may in fact be different. Schematically, the proposed mechanism is as follows. During the cycle, the stable state of the catalytic center, before an additional electron is supplied to the system, is such that one of the metal centers is formally oxidized, e.g., Fe3+H2O or Cu2+H2O. This state is established in a previous step of the cycle, when a "chemical" proton is accepted by one of the hydroxy ligands of the binuclear center. (Here and below for clarity, we distinguish protons that participate in oxygen reduction, which we denote "chemical" protons, from the protons that are pumped through the membrane, which we denote "pumped" protons.) In this state, His291, the so-called proton loading site (PLS), is deprotonated, and Glu242 is protonated. The following steps are involved: (1 and 2) An electron is supplied to the system via cyt c and CuA, which is transferred to heme and then to the heme a3-CuB binuclear center. One of the metal ions is reduced, and the overall charge of the binuclear center (BNC) becomes one charge unit more negative. The driving force of pure electron transfer is about 20 meV.158 (3) In response to the increased negative charge of the binuclear center, the proton from Glu242 now has a driving force to move closer to the binuclear center. There are two pathways leading from Glu242 to two possible sites: one is the BNC itself, and the other is His291sthe PLS. The assumption (gating) is made that the rate of proton transfer to His291 is much higher than that to the binuclear center. Therefore, the protonation of His291 occurs before that of the binuclear center. The fast proton transfer from Glu242 to His291 occurs by the Grotthuss mechanism via Arg438 and PRDa3. The driving force for this transfer is about 100 meV.158 With this transition, His291 becomes protonated. (4) In this step Glu 242 is reprotonated. (5) Now the second, chemical proton is transferred to the binuclear center, using the second path connecting Glu242 and the catalytic center of CcO. The driving force for this transition is about 250 meV.158. A significant driving force, despite the presence of the proton on a nearby His291, is due to formation of a water molecule in the BNCs. This is the main source of energy in the process. (Shang, L., 2017)

This transition occurs after the first proton has moved to the His291 proton loading site. (6) The previously formed state has two additional protons present in two closely located sites (the PLS (His291) and the BNC) but only one additional electron residing on one of the metal ions of the BNC. This state is metastable because of significant proton repulsion. The state is stabilized, therefore, by the expulsion of the proton from the His291 PLS site. The additional energy of stabilization gained is about 250 meV. The expulsion of the proton from His291 from a state in which one of the metal centers is formally oxidized is predicted by electrostatic calculations, which show that, in this redox state, His291 has to be deprotonated. (7) Glu 242 is reprotonated again, and one turnover of the cycle is complete. The former state is stable until the next electron is passed through the system. This last step of reprotonation of Glu242 may be correlated with the expulsion from His291 and/or subsequent rereduction of heme a in the next turnover in the cycle. Each time a proton is pumped, in a single turnover of the pump, a stable state is formed until the next electron is injected into the system. Thus, for each electron passing through the chain, there is one pumped proton. For each oxygen molecule, four electrons are required to form two water molecules; therefore, a maximum of four protons can be pumped. (Kobatake T., 2010)

The above model is based on the energetics of the system described originally in ref 158 (and subsequent studies), and a key kinetic assumption is that, upon reduction of the BNC, the first (pumping) proton is transferred to the PLS of the pump, and later the second, chemical proton, arrives at the BNC. The first proton transfer is fast; however, it leads to a state (proton on the PLS) that is not most favorable energetically. The most energetically favorable state (proton on OH- in the BNC) is achieved by the second, slow proton transfer to the BNC. Since the PLS and BNC sites are closely located, due to electrostatic repulsion, the two protons cannot coexist, and the first proton is expelled for the sake of achieving an energetically more stable state. Since both chemical and pumping protons are derived from the same source (protonated Glu242; one or two chemical protons can possibly come via the K-channel) and due to the special arrangement of the two channels leading to the PLS and to the BNC, the transfer of the second chemical proton blocks the return (back transfer) of the first proton to Glu242.(Ookubo Y., 2012)

CONCLUSION AND FUTURE SCOPE

Transition metal-catalyzed cross-coupling is currently considered the first choice in the synthesis of biaryls. The key step in this process is the reductive elimination of two aryl ligands on the transition metal center. In contrast, our biaryl synthesis totally differs from catalytic cross-coupling in terms of the reaction mechanism and reaction scope. By transiently connecting two aryl groups with a heteroatom–heteroatom linker, sigmatropic rearrangement. Thus, the solvent reorganization energy can be estimated from the experimental measurement of the solvent reorganization energy for the corresponding ET reaction. From the theoretical standpoint, the PCET solvent reorganization energy can be calculated with dielectric continuum models or molecular dynamics simulations, analogous to the methods used for ET.

REFERENCES

Atherton, S. J., & Harriman, A. (1993). Time-resolved studies of intramolecular photoinduced electron transfer in donor-acceptor systems. *Journal of the American Chemical Society*, *115*(5), 1816–1822. DOI: 10.1021/ja00058a020

Benderskii, V. A., & Grebenshchikov, S. Y. (1994). Solvation dynamics and electron transfer in liquids. *Journal of Electroanalytical Chemistry (Lausanne, Switzerland)*, *375*(1–2), 29–44. DOI: 10.1016/0022-0728(94)03471-1

Binstead, R. A., & Meyer, T. J. (1992). Thermodynamics and kinetics of transition-metal oxidation reactions in solution. *Journal of the American Chemical Society*, *114*(1), 173–186. DOI: 10.1021/ja00027a022

Blair, D. F., Ellis, W. R.Jr, Wang, H., Gray, H. B., & Chan, S. I. (1986). Electron transfer in cytochrome c oxidase. *The Journal of Biological Chemistry*, *261*(25), 11524–11537. DOI: 10.1016/S0021-9258(18)67749-1

Bögerhausen, O., Haumann, MJunge, W. (1996). Photosynthetic oxygen evolution: H/D isotope effects and the coupling between electron and proton transfer during transitions S2⟹3 and S3⟹S4→S0. *Berichte der Bunsengesellschaft fur Physikalische Chemie*, *100*(12), 1987–1992. 10.1002/bbpc.19961001210

Brudvig, G. W., Holden Thorp, H., & Crabtree, R. H. (1991). Probing the mechanism of water osidation in photosystem II. *Accounts of Chemical Research*, *24*(10), 311–316.

Eberhart, A. J., & Procter, D. J. (2011). Nucleophilic ortho allylation of aryl and heteroaryl sulfoxides. *Organic Letters*, *13*(22), 5882–5885. DOI: 10.1021/ol202483e

Eberhart, A. J., & Procter, D. J. (2013). Nucleophilic ortho-propargylation of aryl sulfoxides: An interrupted Pummerer/allenyl thio-Claisen rearrangement sequence. *Angewandte Chemie International Edition*, *52*(15), 4008–4011. DOI: 10.1002/anie.201209544

Eberhart, A. J., & Procter, D. J. (2013). Nucleophilic ortho-allylation of pyrroles and pyrazoles: An accelerated Pummerer/thio-Claisen rearrangement sequence. *Organic Letters*, *15*(16), 3994–3997. DOI: 10.1021/ol4016734

Eberhart, A. J., & Procter, D. J. (2015). Sulfoxide-directed metal-free ortho-propargylation of aromatics and heteroaromatics. *Chemistry (Weinheim an der Bergstrasse, Germany)*, *21*(19), 7428–7434. DOI: 10.1002/chem.201500254

Felderhoff, M., Steller, I., Reyes-Arellano, A., Boese, R., & Sustmann, R. (1996). Synthesis and structures of diazomethane-derived cyclopropanes. *Advanced Materials*, *8*(5), 402–405. DOI: 10.1002/adma.19960080507

Graige, M. S., Paddock, M. L., Bruce, J. M., Feher, G., & Okamura, M. Y. (1996). Mechanism of proton-coupled electron transfer for quinone (Q_B) reduction in reaction centers of *Rhodobacter sphaeroides.Journal of the American Chemical Society*, *118*(38), 9005–9016. PMC+1PMC+1 https://doi.org/10.1021/ja960056m

Gust, D., Barrett, D., Harding, L. O., Moore, A. L., Moore, T. A., & Vermaas, W. F. J. (1988). Photochemical energy storage in reaction center model systems. *Journal of the American Chemical Society*, *110*(1), 321–323. DOI: 10.1021/ja00208a048

Gust, D., & Moore, A. L. (1993). Mimicking photosynthetic solar energy transduction. *Accounts of Chemical Research*, *26*(4), 198–205. DOI: 10.1021/ar00029a005

Gust, D., Moore, T. A., Moore, A. L., Gao, F., & Luttrull, D. K.. (1991). Photochemical charge separation in biomimetic molecular triads. *Journal of the American Chemical Society*, *113*(10), 3638–3649. DOI: 10.1021/ja00009a015

He, Z., & Procter, D. J. (2019). Pummerer chemistry of benzothiophene sulfoxides: Metal-free alkylation and arylation of benzothiophenes. *Phosphorus, Sulfur, and Silicon and the Related Elements*, *194*(7), 669–677. DOI: 10.1080/10426507.2019.1619603

Higuchi, K., & Tayu, M. (2021). The interrupted Pummerer reaction: Design of sulfoxides and their utility in organic synthesis. *Heterocycles*, *102*(3), 783–824. DOI: 10.3987/COM-20-14040

Hoganson, C. W., & Babcock, G. T. (1997). A metalloradical mechanism for the generation of oxygen from water in photosynthesis. *Science*, *277*(5334), 1953–1956. PMC https://doi.org/10.1126/science.277.5334.1953

Huang, X., Klimczyk, S., & Maulide, N. (2012). Charge-accelerated sulfonium [3,3]-sigmatropic rearrangements. *Synthesis*, *44*(2), 175–183. DOI: 10.1055/s-0031-1289807

Huang, X., & Maulide, N. (2011). Sulfoxide-mediated α-arylation of carbonyl compounds. *Journal of the American Chemical Society*, *133*(21), 8510–8513. DOI: 10.1021/ja202591s

Huang, X., & Maulide, N. (2013). Sulfur(IV)-mediated transformations: From ylide transfer to metal-free arylation of carbonyl compounds. *Journal of the American Chemical Society*, *135*(20), 7312–7323. DOI: 10.1021/ja402710j

Hung, S.-C., Macpherson, A. N., Lin, S., Liddell, P. A., & Seely, G. R. (1995). Photocurrent generation by molecular triads in photoelectrochemical cells. *Journal of the American Chemical Society*, *117*(6), 1657–1658. DOI: 10.1021/ja00109a034

Kaiser, D., Oost, R., Neuhaus, J., & Maulide, N. (2019). Bond-forming and –breaking reactions at sulfur (IV): Sulfoxides, sulfonium salts, sulfur ylides, and sulfinate salts. *Chemical Reviews*, *119*(13), 8701–8780. DOI: 10.1021/acs.chemrev.9b00134

Kitagawa, T., & Ozawa, T. (1997). Resonance Raman spectroscopy of metalloporphyrins and heme proteins. In Lippard, S. J. (Ed.), *Progress in Inorganic Chemistry* (Vol. 45, pp. 431–479). Wiley., DOI: 10.1002/9780470166413.ch8

Kobatake, T., & Oshima, K. (2010). Reaction of 2-(2,2,2-trifluoroethylidene)-1,3-dithiane 1-oxide with ketones under Pummerer conditions and its application to the synthesis of trifluoromethyl-substituted five-membered heteroaromatics. *Angewandte Chemie International Edition, 49*(13), 2340–2343. DOI: 10.1002/anie.200907016

Kobatake, T., & Oshima, K. (2010). Synthesis of 3-trifluoromethylbenzo[b]furans from phenols via direct ortho functionalization by extended Pummerer reaction. *Journal of the American Chemical Society, 132*(33), 11838–11840. DOI: 10.1021/ja104317a

Krishtalik, L. I. (1970). Electrochemical mechanisms and kinetics of electron transfer reactions. In Delahay, P., & Tobias, C. W. (Eds.), *Advances in Electrochemistry and Electrochemical Engineering* (Vol. 7, pp. 283–339). Interscience.

Lavergne, J., & Junge, W. (1993). Proton release during the redox cycle of the water oxidase. *Photosynthesis Research, 38*(3), 279–296. Groningen Research Portal+4SURFconext+4PMC+4 https://doi.org/10.1007/BF00046752

Levich, V. G. (1970). Physicochemical hydrodynamics. In H. Henderson & W. Yost (Eds.), *Physical Chemistry—An Advanced Treatise* (Vol. 9B, pp. 985–1074). Academic Press.

Meyer, B., Schlodder, E., Dekker, J. P., & Witt, H. T. (1989). O_2 evolution and Chl a^+II (P680$^+$) reduction in photosystem II: A comparative study of the kinetics in spinach thylakoids and in the isolated D1/D2/cytochrome b-559 complex. *Biochimica et Biophysica Acta (BBA) -. Bioenergetics, 974*(1), 36–43. PMC https://doi.org/10.1016/S0005-2728(89)80163-X

Ookubo, Y., & Osuka, A. (2012). Synthesis of a library of fluorescent 2-aryl-3-trifluoromethylnaphthofurans from naphthols by using a sequential Pummerer annulation/cross-coupling strategy and their photophysical properties. *Chemistry (Weinheim an der Bergstrasse, Germany), 18*(40), 12690–12697. DOI: 10.1002/chem.201201930

Pecoraro, V. L. (Ed.). (1992). *Manganese redox enzymes*. VCH Publishers.

Pulis, A. P., & Procter, D. J. (2016). C–H coupling reactions directed by sulfoxides: Teaching an old functional group new tricks. *Angewandte Chemie International Edition, 55*(33), 9842–9860. DOI: 10.1002/anie.201512854

Roecker, L., & Meyer, T. J. (1987). Hydride transfer in the oxidation of alcohols by [(bpy)2(py)Ru(Q)]2+. A kH/kD kinetic isotope effect of 50. *Journal of the American Chemical Society, 109*(3), 746–754. 10.1021/ja00237a019

Shafir, A. (2016). The emergence of sulfoxide and iodonio-based redox arylation as a synthetic tool. *Tetrahedron Letters, 57*(24), 2673–2682. DOI: 10.1016/j.tetlet.2016.04.036

Shafirovich, V. Y., Courtney, S. H., Ya, N. Q., & Geacintov, N. E. (1995). Reactions of superoxide radical anion with aromatic compounds in aprotic media. *Journal of the American Chemical Society, 117*(17), 4920–4929. DOI: 10.1021/ja00121a016

Shang, L., & Zhang, L., (2017). Redox-neutral α-arylation of alkyl nitriles with aryl sulfoxides: A rapid electrophilic rearrangement. *Journal of the American Chemical Society, 139*(12), 4211–4217. DOI: 10.1021/jacs.7b01204

Tommos, C., Tang, X.-S., Warncke, K., Hoganson, C. W., & Styring, S., McCracken, J., Diner, B. ABabcock, G. T. (1995). Spin-Density Distribution, Conformation, and Hydrogen Bonding of the Redox-Active Tyrosine YZ in Photosystem II from Multiple-Electron Magnetic-Resonance Spectroscopies: Implications for Photosynthetic Oxygen Evolution. *Journal of the American Chemical Society, 117*(39), 10325–10335.

Vincent, J. B., & Christou, G. (1989). Higher Oxidation State Manganese Biomolecules. *Advances in Inorganic Chemistry, 33*, 197–257.

Yanagi, T., & Yorimitsu, H. (2018). Recent development of ortho-C–H functionalization of aryl sulfoxides through [3,3] sigmatropic rearrangement. *Tetrahedron Letters, 59*(29), 2951–2959. DOI: 10.1016/j.tetlet.2018.06.011

Yoshida, S., & Oshima, K. (2009). 2-(2,2,2-Trifluoroethylidene)-1,3-dithiane monoxide as a trifluoromethylketene equivalent. *Organic Letters, 11*(10), 2185–2188. DOI: 10.1021/ol900717d

Zhang, L., & Peng, B. (2019). [3,3]- and [5,5]-Sigmatropic rearrangements of aryl sulfoxides using an 'assembly/deprotonation' technology. *Synlett*, *30*(20), 2203–2208. DOI: 10.1055/s-0037-1610257

Chapter 4
Reactive Intermediates: Carbocations, Carbanions, Free Radicals, and Carbenes

S. Anitha
https://orcid.org/0000-0002-6735-2831
Sri Sairam Engineering College, Chennai, India

M. Kannan
Ramakrishna Mission Vivekananda College, Chennai, India

ABSTRACT

Reactive intermediates are chemical species that are short lived and unstable molecules formed during the overall reaction pathway. These short-lived intermediates are formed via homolytic or heterolytic cleavage of bond. The transient nature of intermediates makes them special. During the reaction they form and are nearly instantly consumed, typically in a matter of seconds. Reactive intermediates are essential for comprehending the reaction mechanism. They are effective catalysts for chemical change because of their instability; they "want" to transform into more stable forms and hence propel the reaction to its conclusion. These transitory entities mediate a series of chemical reactions necessary for the production of dyes and dye intermediates. They offer the key steps in the synthetic pathways that enable us to efficiently and more sustainably manufacture complex compounds.

DOI: 10.4018/979-8-3693-6473-4.ch004

INTRODUCTION

The most rewarding and fascinating area of chemical research that has flourished as a result of instrumentation and related technological advancements is the study of reactive intermediates. Reactive intermediates are short-lived, highly reactive, high-energy molecules that, when produced in a chemical reaction, immediately convert into stable molecules (Smith & March, 2006). They are stored separately in some circumstances. A great deal of reactions occurs through the generation of intermediates (Vajira, 2013). They are transient and get involved in the reaction as soon as they form. It is not possible to isolate and store these intermediates under normal conditions but only detectable using quick spectroscopy techniques. Sometime, experimentation is required to infer its presence. Typically, this entails adjusting the temperature or concentration of the reaction as well as using spectroscopy, chemical thermodynamics, and kinetics approaches. Some of the key intermediates in a wide range of chemical reactions are Carbocations, carbanions, free radicals, and carbenes.

Understanding the chemistry of reactive intermediates is crucial to comprehending organic chemistry. Only with this knowledge, it is feasible to predict the reactions and products that will occur via the reactive intermediates. Otherwise, it would be hard to master organic chemistry, which has a very wide range of applications. This chapter is aimed at guiding the audience with essential information related to this topic.

This chapter aims to elaborate on generation of reactive intermediates (carbocation, carbanion, free radicals and carbene). It aims to take a deep dive into fascinating topics in chemistry reactions involving reactive intermediates and explore their structure, their distinct characteristics, various kinds, and vital function in various processes. For chemistry lovers, students, and professionals alike, this article delivers an accessible and captivating introduction to the fascinating world of reactive intermediates. It concentrates mainly on their structure, stability, generation and its application in various fields. It concentrates mainly on their structure, stability, generation and its application in various fields. It is capped by a large problem set for the readers to explore more. This chapter is staunched to explain about reactive intermediates role in organic chemistry.

CARBOCATIONS

Carbocation, also known as carbonium ion, is an organic species in which each carbon atom has three bonds and a positive charge. It is an extremely reactive, high-energy, unstable, and electron-deficient species. The reason for their high energy and reactivity is that the carbon carbocations don't follow the octet rule because they only have six electrons in their outer shell and tend to complete its octet. These reactive intermediates are generated via heterolytic fission of a covalent bond involving a carbon atom and are often encountered as transient intermediates in large number of organic reactions. Especially in nucleophilic substitution reaction ($S_N{}^1$) and elimination (E1) reactions of alkyl halides (Singh and Yadav, 2010).

Structure

The carbon atom of carbocation is sp^2 hybridised in which the unhybridized p-orbital is devoid of any electron. The three bonding orbital form three sigma bonds whereas, the remaining unhybridized p-orbital is vacant. This vacant unhybridized p-orbital makes the carbon atom electron deficient and gives it a positive charge. Therefore, carbocation has a trigonal planar (flat) structure having all the three sigma bonds in one plane and the bond angles of 120° between them.

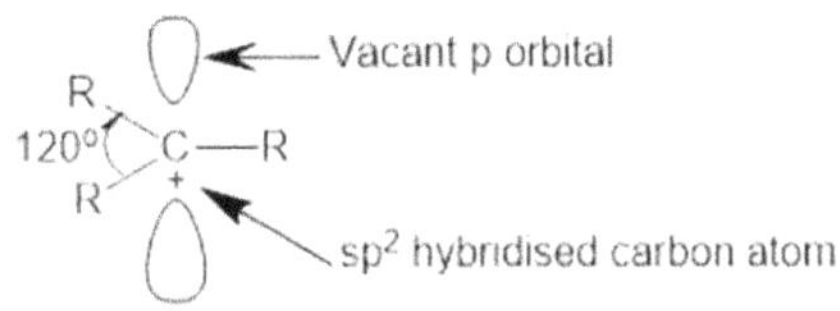

Carbocations are Classified Into

a) Alkyl carbocation
b) Aryl carbocation
c) Allylic carbonations
d) Benzylic carbocations

e) Vinyl carbocations

Alkyl Carbocation. Alkyl carbocation contains alkyl groups attached to the +ve carbon atom.

It is classified as

A primary carbocation (1° carbocation): When one carbon group is attached to the carbon that has the positive charge, it is referred to as a primary carbocation.

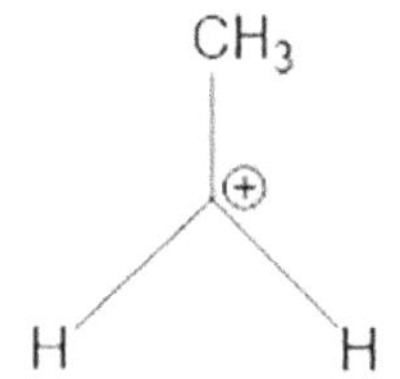

A secondary carbocation (2° carbocation): When two carbons are attached to the carbon that has the positive charge, it is referred to as a secondary (2°) carbocation.

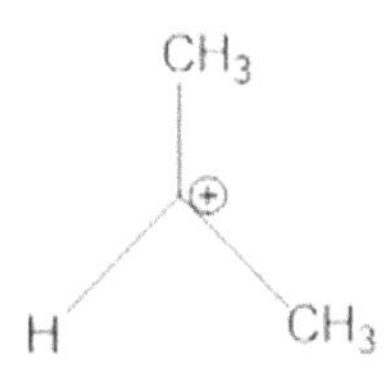

A tertiary carbocation (3° carbocation): When three carbons are attached to the carbon that has the positive charge, it is referred to as a tertiary (3°) carbocation.

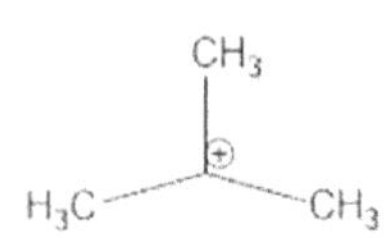

Aryl Carbocations. An aryl carbocation is a carbocation, if the carbon bearing positive charge belongs to a benzene ring. It is sp^2 hybridized, highly unstable as the positive charge on the double bonded carbon is highly electronegative and destabilize it. In aryl carbocations, since the p electron clouds are perpendicular to the vacant p orbital of the carbocation it doesn't have resonance stabilization.

Allylic Carbocations. The cation at the allylic carbon i.e., carbon attached directly or indirectly to a double bond is named as allylic carbocations. The allylic carbon atom carries a positive charge. This carbocations forms intermediate in many organic reactions. The allylic carbon atom is sp^3 hybridized as it bonded with $-CH=CH_2$ through a single covalent bond. Further, this allylic carbocations are classified as primary allylic carbocation as the formal charge of +1 is on the primary carbon atom, and in secondary allylic carbocation, the +1 formal charge is distributed on the secondary carbon atom and in tertiary allylic carbocation with +1 charge on a tertiary carbon atom of the cation.

The stability of carbocations determined by steric hindrance and +I effect of alkyl group attached to C^+ of the carbocation. The stability is due to delocalization of electrons on carbon atom. Allyl carbocations have similar stability as secondary carbocations.

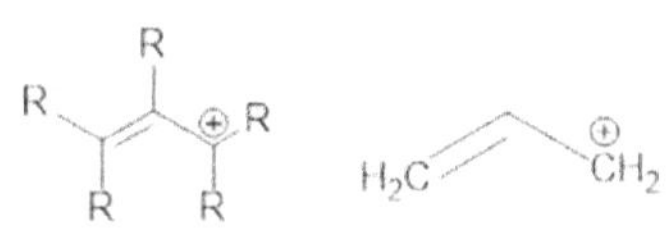

Benzylic Carbocations. A benzylic carbocation is a carbocation, if the carbon bearing positive charge is immediately adjacent to a benzene ring.

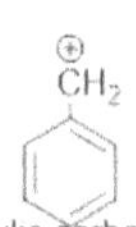

Benzylic carbocation
sp bridised carbocation
less electronegative
less % of s character

Vinyl Carbocations. If the carbon bearing the positive charge is part of an alkene, the carbocation is termed a vinylic carbocation.

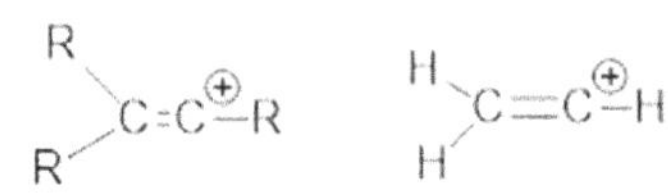

Stability

The stability of carbocation are as follows

3° >2° >1°> CH_3

Factors Affecting the Stability of Carbocations

- Presence of adjacent alkyl groups increases the carbocation stability.
- Presence of adjacent pi bonds that allow delocalization through resonance increases the carbocation stability.
- Presence of adjacent atoms with lone pairs that can fulfill the carbon octet, increases the carbocation stability

The above factors can be explained as shown below

i. Hyperconjugation
ii. Inductive effect
iii. Resonance effect

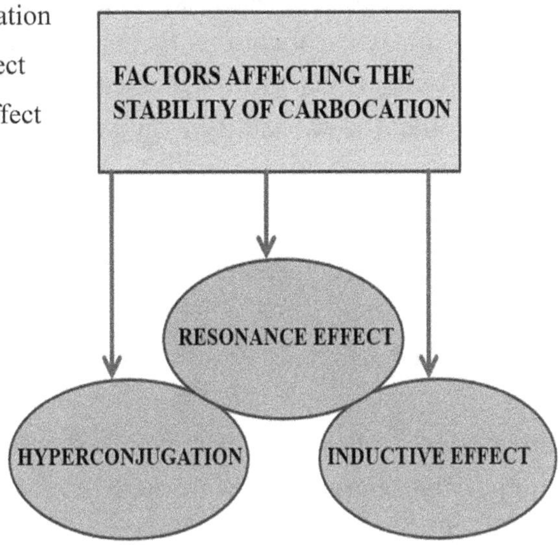

Hyper Conjugation

Hyper conjugation effect is a permanent effect due to electron displacement, involving localization of σ electrons of alkyl C-H bond directly attached to an atom of the unsaturated system or to an atom with an unshared p orbital takes place. It is also named as sigma bond resonance.

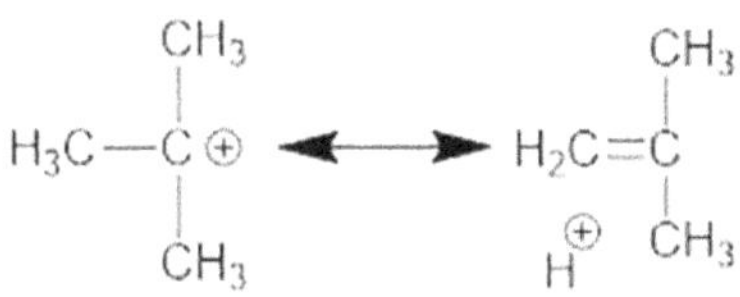

The above figure shows the number of hyper conjugative forms for tertiary carbocations.

Due to hyperconjugation, the electrons in the adjacent alkyl C-H bonds overlaps with the empty p-orbital of the positive charged carbon of carbocation and result in stabilisation. More the number of hyper conjugating structures more will the stability. Tertiary carbocations forms nine hyperconjugation structure, while secondary has six hyperconjugation structure and primary carbocations forms three hyperconjugation structures respectively. It is clear from the above that the tertiary carbocations are more stable than the secondary which in turn are more stable than the primary carbocations.

$3^o > 2^o > 1^o > CH_3$

$R_3C^+ (3^o) > R_2CH^+ (2^o) > RCH_2^+ > (1^o)\ CH_3^+$ (methyl; least stable)

Inductive Effect

Inductive effect is due to the permanent dipole moment caused by the negative or positive charge relayed through the carbon atom due to electron releasing or electron withdrawing species.

Alkyl groups have positive inductive effects (+I). The presence of (+I effect) of alkyl group, which increases the electron density at the positive charge bearing carbon, the tertiary carbocations with three adjacent alkyl group are more stable due to more electron donation which make it more electron rich than secondary and primary carbocations, therefore, the order is tertiary carbocation> secondary carbocation > primary carbocation. i.e., The greater stability of tertiary carbocations is due to the greater neutralisation of positive charge on the carbon by three alkyl group compared to secondary and primary carbocations which as two alkyl and one alkyl group respectively. Thus, tertiary carbocation is more stable according to the inductive effect.

Figure 10. Number of alkyl group increases

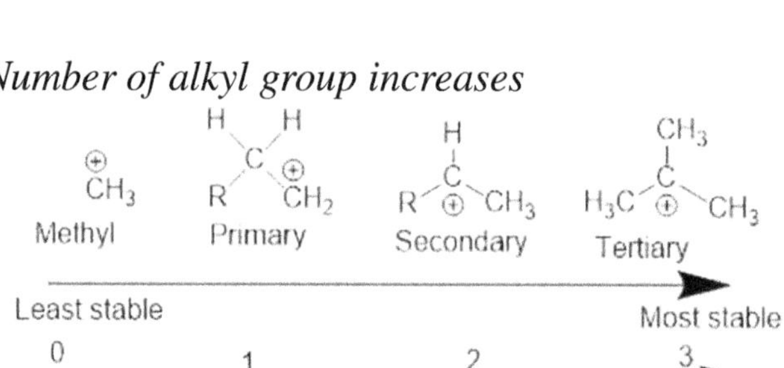

Resonance Effect

Resonance effect is due to the interaction between a lone pair and a pi bond or between two C-C pi bonds between adjacent atoms which results in polarity in a molecule and leading to spreading-out of charges.

The stability of carbocation is also based on resonance effect as it reduces the positive charge density on carbonations making it more stable. Greater the number of resonating structure greater will the stability. Hence the triphenylmethyl cation with nine resonating structure have more stability than diphenyl methyl cation with six resonating structure. The +ve charge of the methyl carbon is delocalized in three phenyl rings, which makes the triphenylmethyl cation extremely stable. A +ve charge is created in two ortho and para positions of every phenyl ring. that is, three structures that resonate. Nine total resonating structures are provided by the triphenylmethyl cation. Thus, delocalisation of pi electron through resonance increases the stability of carbocations.

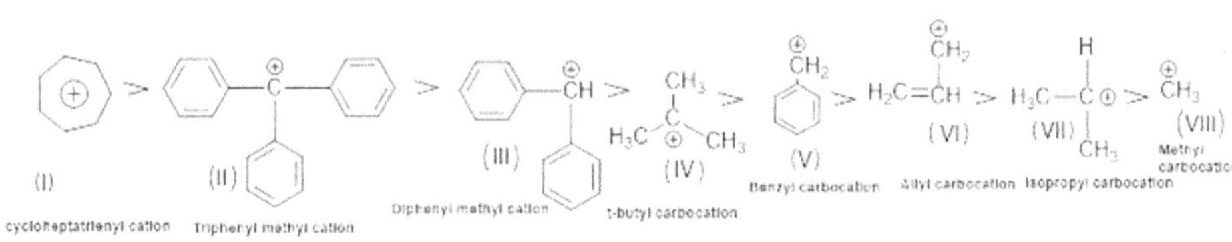

The comparative stability of carbonations is as follows

1. The cycloheptatrienyl cation is an exceptionally stable carbocation which can be easily understood on the basis of resonance. It has seven equivalent resonance forms. In add to this, these carbocations are aromatic in nature and the positive charge is also a part of the aromatic ring which results in complete delocalisation and hence more stable.
2. The stability of triphenyl carbocations is also due to its nine resonance forms with positive charge on the central carbon atom.
3. The number of resonance structure is less for diphenyl methyl carbocation than the triphenylmethyl carbonations hence lesser the stability.

4. Benzyl carbonations has less stability compared to the above three mentioned carbocations, since in this in this case number of resonating structures are less.
5. Positive charge in cycloheptatrienyl cation are present on a part of an aromatic ring. Due to this reason, it is very stable compared to triphenylmethyl, diphenyl methyl and benzyl carbocations in which positive are present on alkyl carbon or on a carbon which is not part of an aromatic ring.
6. Then comes the stability of allyl with only two resonating forms, followed by tertiary > secondary > Primary carbocations whose stability order can be better explained by hyperconjugation and by field effect (Inductive effect) (as already discussed).

Generation of Carbocation

The carbocations is formed by

Protonation of Alkenes and Carbonyl Compounds. The following are the general reactions employed for the generation of carbocations from alkenes and carbonyl compounds

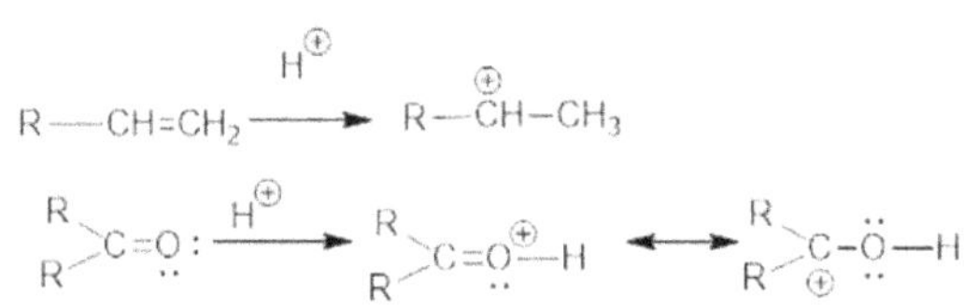

Direct Ionization. In direct ionization a leaving group attached to a carbon atom leaves with its pair of electrons resulting in a carbocation formation. It may be a S_N1 or E1 reaction depending upon whether it leads to a substitution or elimination product.

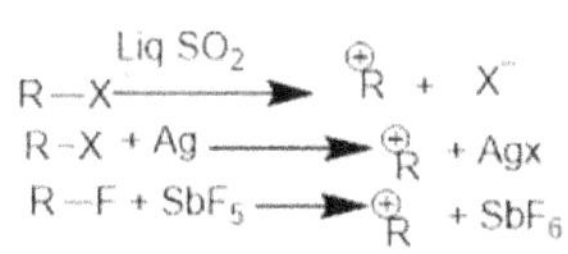

Deamination of Amines in Presence of Nitrous Acid (HNO2). Deamination of amines in presence of nitrous acid provides a useful test for distinguishing primary, secondary and tertiary amines.

- Primary amines + HNO_2 → Nitrogen Gas Evolution
- 2°-Amines + HNO_2 → N-Nitrosamine
- 3°-Amines + HNO_2 → Ammonium Salt Formation

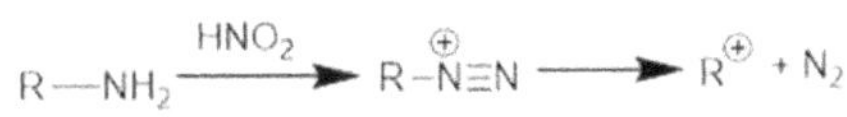

The above reactions proceed via carbocation formation.

Protonation of Alcohols Followed by Dehydration. Protonation of alcohol by adding a protonic acid result in the formation of a better leaving group, resulting in C-O bond breakage generating a carbocation. Further dehydration occurs in presence of a base resulting in alkene formation.

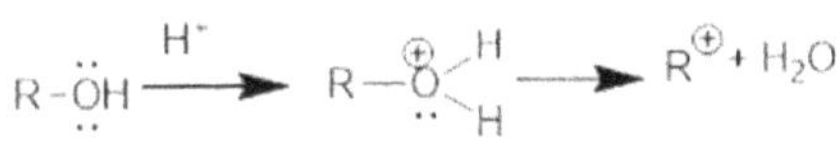

REACTION INVOLVING CARBOCATION INTERMEDIATE

Carbon-Carbon Bond Formation

Friedel Craft Reaction (Alkylation and Acylation). An electrophilic aromatic substitution known as the Friedel-Crafts reaction occurs when a carbocation attacks the pi bond of an aromatic ring, replacing one of the aromatic protons with an alkyl group.

In Friedel-Crafts alkylation and acylation, the electrophilic attack on the π system that forms carbon-carbon bonds is an essential step. Because a π bond replaced by a strong σ bond, the interaction between a carbocation and an alkene to generate a new carbon-carbon bond is both thermodynamically and kinetically favourable.

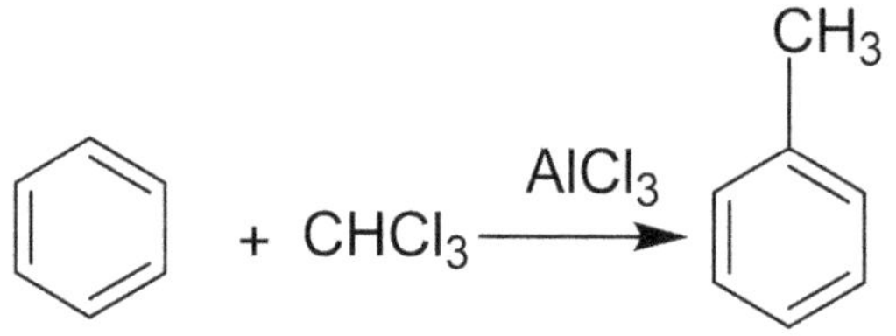

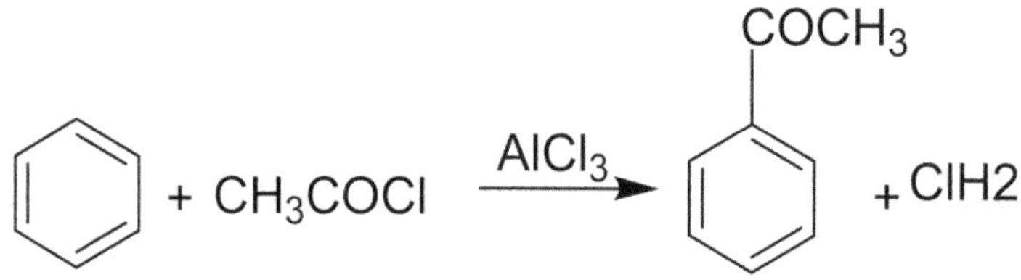

Alkylation of Silyl Enol Ethers. Alkylation of silyl enol ethers with tertiary alkyl halides, forms stable carbocations in the presence of Lewis acids like $TiCl_4$ or $SnCl_4$. This reaction offers a way to introduce tertiary alkyl groups (α) to a carbonyl, a transformation that is not possible with base-catalyzed alkylation due to the strong elimination tendency of tertiary halides.

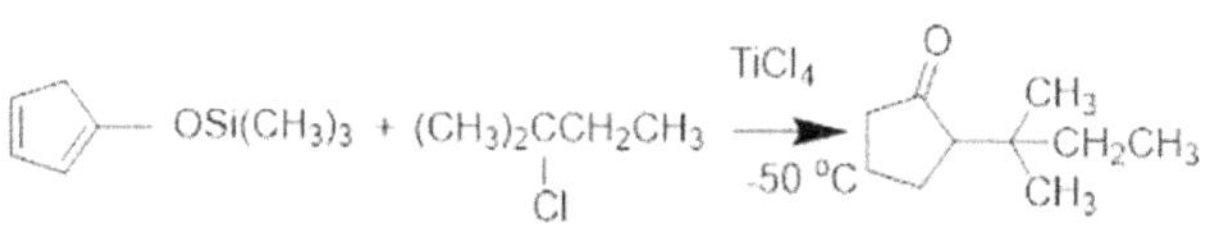

Rearrangement Reactions Involving Carbocations

Pinacol-Pinacolone Rearrangement. Reaction in organic chemistry to convert 1,2diol to a carbonyl compound. It occurs by 1,2 rearrangements (Ahluwalia and Parashar, 2011).

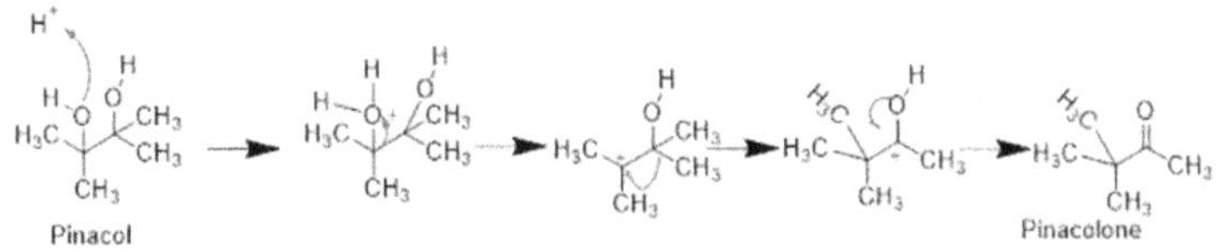

Pinacol an organic compound with two hydroxyl group attached to vicinal carbon atom in acidic medium which undergoes 1,2 shift to form pinacolone through a stable tertiary carbocation intermediate.

Wagner–Meerwein Rearrangement. Organic reaction involving 1,2 rearrangements in which a hydrogen, alkyl or aryl group migrates to adjacent carbon atom. In this rearrangement reaction an alcohol is converted to olefin using acid catalyst via a carbocation intermediate.

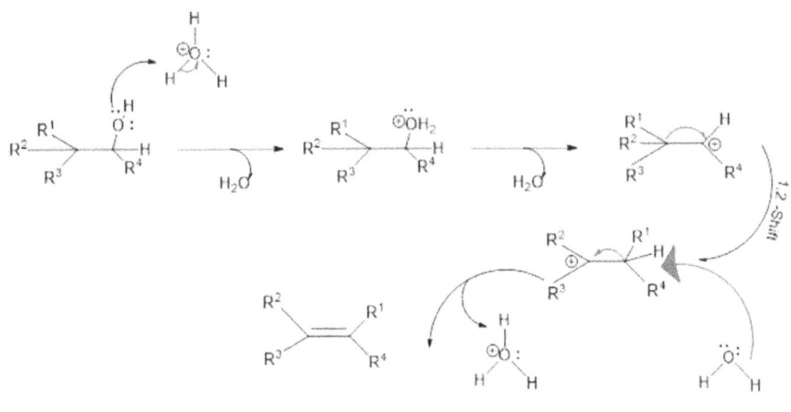

Elimination Reactions (El)

Alkenes from Alkyl Halides and Alcohols. Alkenes can be obtained from alkyl halide such as bromo and iodo and less commonly, chloro derivatives (Normon and Coxon, 1993).

Haloalkanes on heating with alcoholic KOH loses one molecule of hydrogen halide to give alkene via carbocation intermediate.

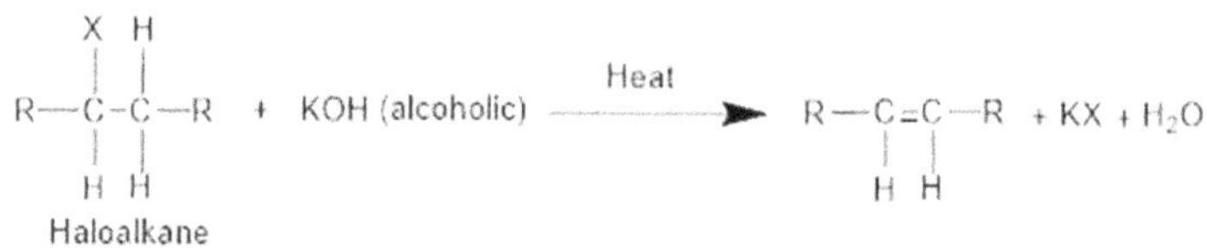

CARBANION

A member of a class of organic compound with a negative charge on trivalent carbon atom formed by removal of positively charged atoms or group of atoms (Hornback, 2005). It is an electron rich intermediate with eight electrons in its octet. Thus, a carbanion is not electron deficient. The presence of negative charge on a carbon atom gives good nucleophilic properties. It is chief intermediates in many organic reactions. It acts a good nucleophile in many organic reactions like substitution and addition reactions. An electron-donating groups bonded to the anionic center of carbanion destabilizes it. Carbanions plays important role in organic chemistry because they can act as nucleophiles, which means they can donate electrons to other molecules.

Structure

Carbanion is sp^3 hybridized and expected to have a tetrahedral structure but since one of the hybrid orbitals contains an unshared pair of electrons its structure becomes pyramidal geometry. As carbanion have one non-bonding and three bonding electron pair in four sp3 hybridized orbital. - The non-bonding electron pair repels the bonding pair more than the bonding pairs repel each other and because of this, there is reduction of the angle between bonding pairs to a value slightly lesser than the tetrahedral value of 109.5°. - The configuration of simple carbanions such as the methyl anion thus appears to be pyramidal just like that of ammonia with which the methyl anion is isoelectronic.

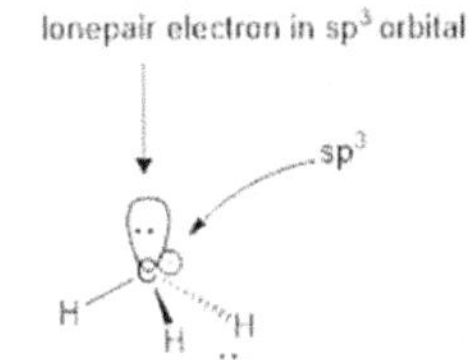

Methyl anion ($\ddot{C}H_3$)

Carbanions behave as Lewis's acids. It has high electronegativity and highly reactive in nature.

Stability of Carbanions

Carbanions stability is increased by factors that tend to reduce the carbanion's negative charge density.

The following factors influence the carbanion stability

a) Electronegativity of carbanionic carbon
b) Hybridization of carbanionic carbon
c) Resonance effect
d) Inductive Effect

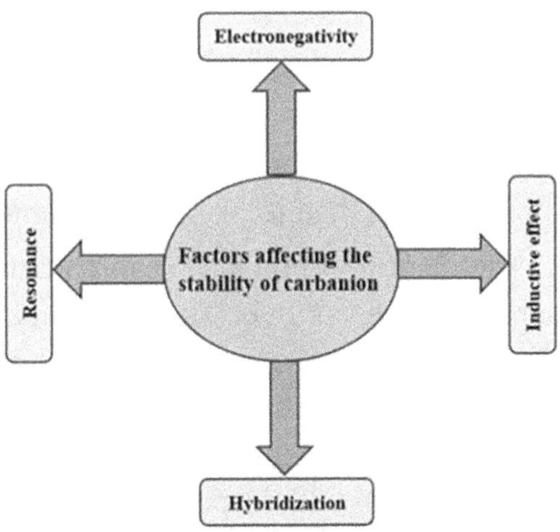

Electronegativity of Carbanionic Carbon. The electronegativity of carbanionic carbon is directly proportional to its s-character and intern with its stability.

(i) Increase in % s-character
(ii) Increase in stability

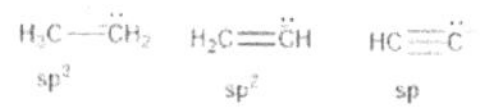

Electron Withdrawing Groups (EWG) such as carbonyl, nitro, and sulfone increase the stability of the carbanion increases by reducing the carbanion's negative charge density.

Thus, as the number of EWG increases as illustrated in the example below, the stabilisation of negative charge on carbanion increases due to dispersal of negative charge on the adjacent EWG atom.

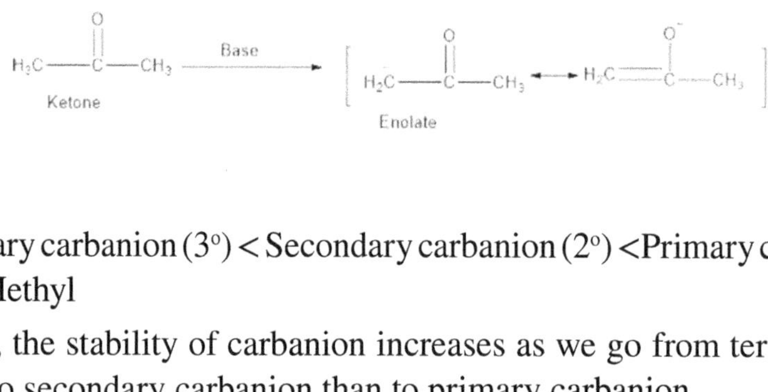

Tertiary carbanion (3°) < Secondary carbanion (2°) <Primary carbanion (1 °) <Methyl

Thus, the stability of carbanion increases as we go from tertiary carbanion to secondary carbanion than to primary carbanion.

Hybridization. Another important factor in carbanions' stability is hybridization. Around the core, sp^3 hybridized carbon atom, a carbanion has three bonding electron pairs and one unshared pair of electrons with a pyramidal structure. Three of the four sp^3 bonds, the fourth is a non-bonding orbital. Carbanions that experience significant resonance stabilization are in the sp2 hybrid state and planar with respect to the delocalized electronic system. On the other hand, a carbanion stabilized by conjugation with substituents has a very different geometry Carbanions that experience significant resonance stabilization are in the sp2 hybrid state and planar with respect to the delocalized electronic system. In simpler the stability of carbanion is proportional to electronegativity of carbanionic carbon and also to the % s -character of carbanionic carbon.

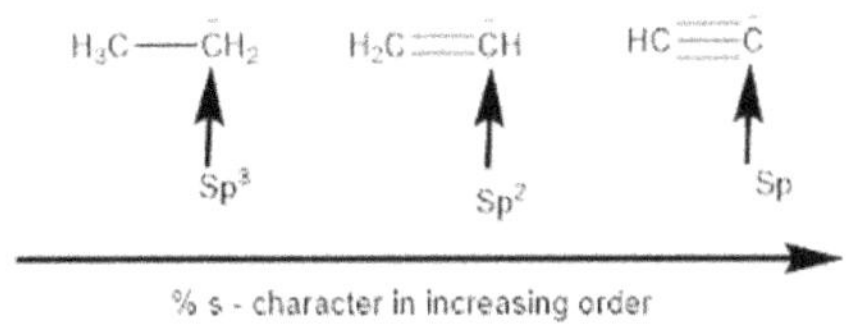

Inductive Effect. Similar to carbocation, the stability of carbanion also depends on the inductive effect.

1. With increase in +I effect stability decreases
2. With increase in –I effect stability increases
3. More electron donating group (+I effect) adjacent to the negative charge bearing carbanion lesser will be the stability.

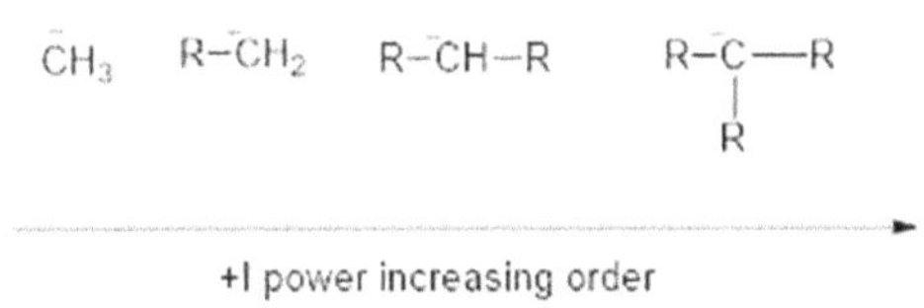

Tertiary carbanion (3°) < Secondary carbanion (2°) <Primary carbanion (1 °) < Methyl

Whereas more the number of electron withdrawing groups (-I effect) adjacent to the negative charge bearing carbanion greater will be the stability.

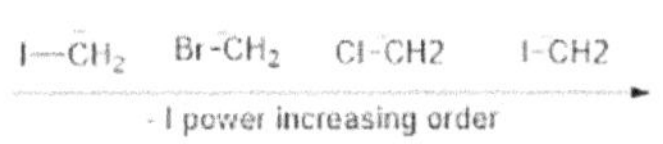

Resonance Effect

Allyl and benzyl carbanions are stabilized by delocalization of negative charge. More the number of resonating structure greater will be the stability.

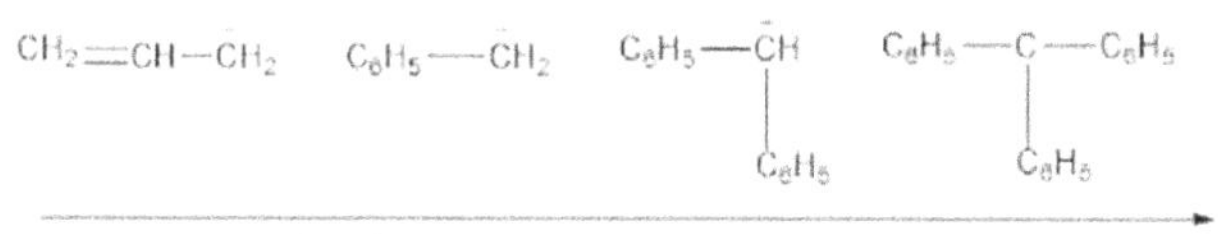

Number of resonating structures is in increasing order

Stability increases

Other Factors Affecting the Stability of Carbanion

1. Carbanion adjacent to Sulphur and phosphorus atom increases the carbanion stability. The delocalization of negative charge of carbanion on to the vacant d- orbital p∏ - d∏ bonding of phosphorus and Sulphur.
2. Presence of >C=O, $-NO_2$ and CN groups on carbanionic carbon by resonance effect stabilize the carbanion.
3. Negative charge present on carbon of aromatic system is known as aromatic carbanions. It is the stable carbanion.

Generation of Carbanion

Some of the suitable method for generation of carbanion involves.

a) Abstraction of protons
b) Nucleophile addition to alkene
c) Organometallic compound formation
d) Decarboxylation

Abstraction of Protons

Formation of an anion by proton abstraction to form Carbanion.

R–H → R^- + H^+

The abstraction of an acidic proton from a base.

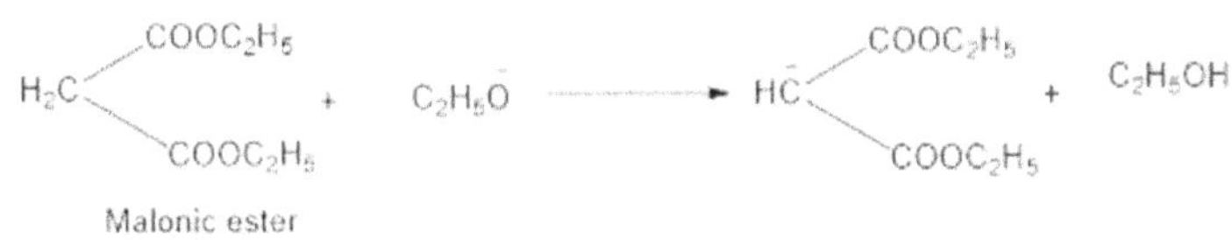

To generate a stable carbanion, organic compounds which possess a labile or acidic hydrogen lose a proton in the presence of weak or a strong bases which abstracts the acidic proton.

Nucleophile Addition to Alkene

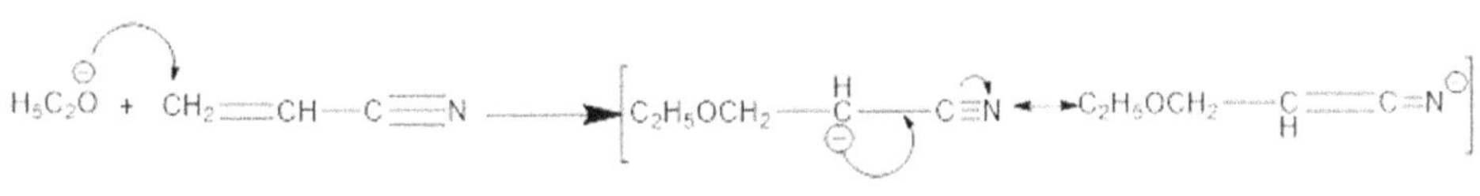

The attack of nucleophiles on any one carbon atom of α,β-Unsaturated results in carbanions forming a negative charge on the left out carbon.

Abstraction of Terminal Hydrogen from Acetylene

Organometallic Compounds Formation

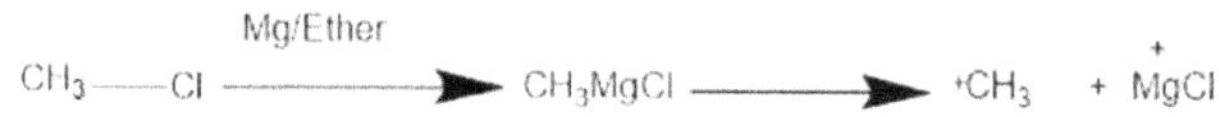

REACTIONS OF CARBANIONS

Aldol Condensation

It is a organic reaction that occurs in aldehydes having α-hydrogen with a dilute base to give β-hydroxy aldehydes called aldols. This reaction is most commonly known as aldol condensation This acidic property of α-hydrogen in aldehydes and ketones is due to the high electron-withdrawing capacity of the carbonyl group and resonance stabilization of the conjugate base.

Aldols are formed by removing a α-hydrogen from the base which result in carbanion formation which later undergoes a nucleophilic addition reaction with the carbonyl group present in the second molecule of the aldehyde resulting in the condensation product.

Cannizzaro's Reaction of Aldehydes Without α-H Atoms. A non-enolizable aldehyde (i.e., aldehyde without α-hydrogen) in the presence of concentrated alkali solution, undergo simultaneous oxidation and reduction to form equivalent alcohol and a carboxylic acid is known as Cannizzaro's reaction.

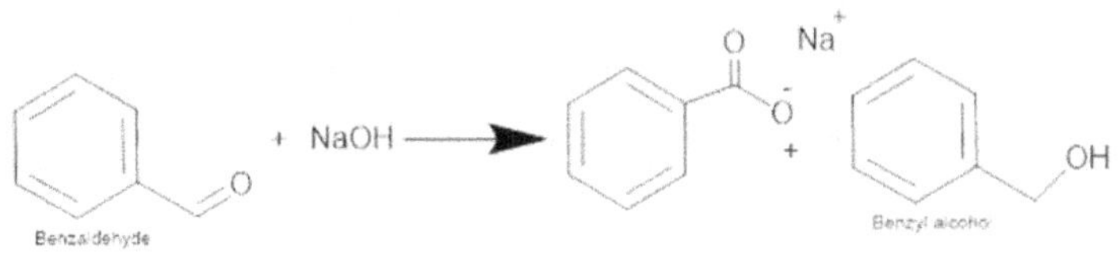

Perkin's Reaction Involves the Formation of Carbanions as Intermediate. When an aromatic aldehyde is heated with excess of anhydride in presence of alkali salt which acts as a base catalyst to speed up the reaction, it produces a β- unsaturated aromatic acid as a byproduct. Carbanion on the carbonyl carbon is the intermediate in the reaction.

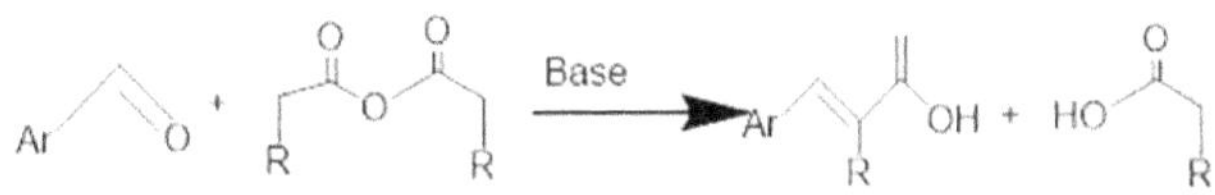

Knoevenagel Reaction Involves the Formation of Carbanions' as Intermediate. Formation of an α, β-unsaturated ketone by a nucleophilic addition between an aldehyde or ketone, and an active hydrogen compound in the presence of a basic catalyst followed by spontaneous dehydration is known as **Knoevenagel reaction**.

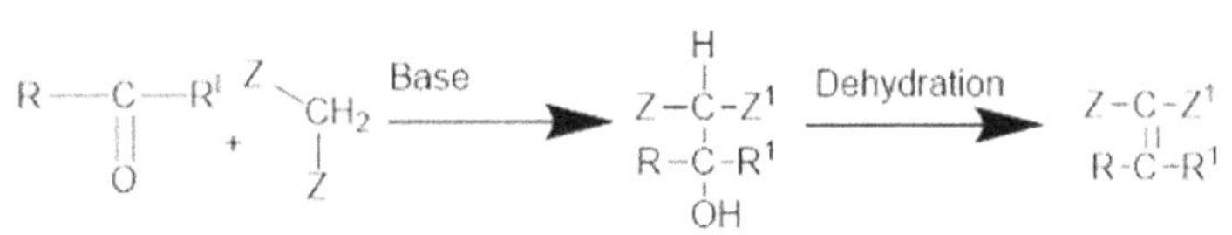

FREE RADICAL

A homolytic cleavage of a covalent bond resulting in carbon with unpaired electron or odd electron i.e., with valence shell containing seven electron only is termed as free radical. This chemical species contains a singly occupied orbital and it is a neutral species and hence tend to be highly reactive. Since the carbon is lack of electron (electron deficient) it acts as an electrophile. Free radicals are generally less stable and react in fraction of seconds with another species (Santos, 2010).

Structure

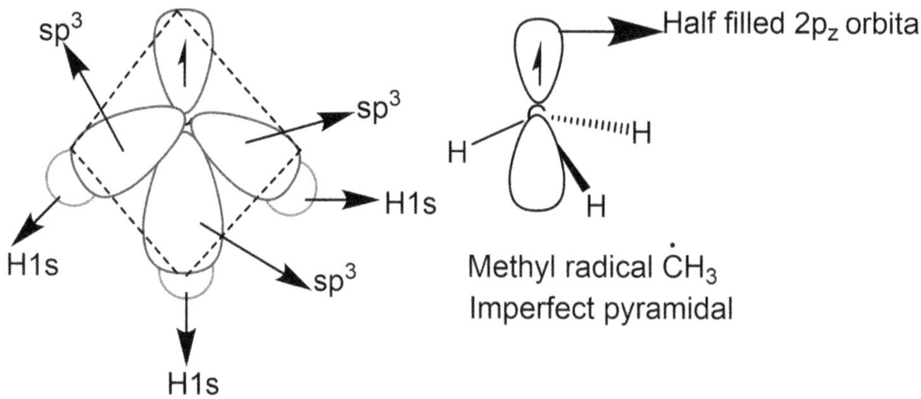

Methyl radical $\dot{C}H_3$
Imperfect pyramidal

Free radicals have different structures depending on the central atom. Free radicals containing carbon and an unpaired electron may also be in one of two hybrid states: either sp² hybridized, which would result in a planar structure with an odd electron in the p orbital, or sp³ hybridized, which would result in a pyramidal structure. Mostly all free radicals have planar geometry; however, the carbon atom carrying the unpaired electron in a bicycle compound free radical cannot acquire a planar geometry, resulting in a pyramidal structure. The reason for this is that although it is not entirely pyramidal due to the lone electron in the 2pz orbital, electron-electron repulsion causes the bonds to bend from trigonal planar geometry.

Stability of Free Radical

The following factors contribute to the stability of Free Radicals

a) Inductive effect
b) Resonance effect
c) Hybridization

d) Adjacent atoms with lone pairs
e) Electronegativity of adjacent atoms
f) Adjacent electron withdrawing group
g) Size of the atom

Figure 38. Factors affecting the stability of free radicals

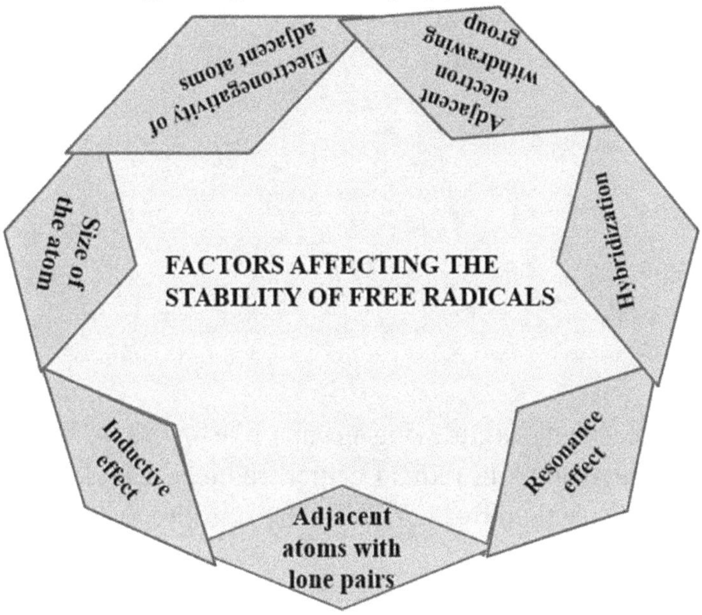

Inductive Effect

Stability of free radicals are influenced by both the +I effect and -I effect, which is the ability of substituents to donate or withdraw electrons through a Sigma bond.

The strength of inductive effect on the stability of free radical depends upon the following factors.

1. Nature of substituents (Electron donating groups, Electron withdrawing groups)
2. Distance between substituent and radical center.

3. Its molecular geometry.

Free radicals which have less electron in the octet are mostly stabilized by electron donating groups like $-CH_3$, -OH, $-NH_2$, -SH. More the number of electrons donating alkyl groups attached to the free radical carbon centre more will be the stability of the radical. This is due to the fact that the electron deficiency of the radical is decreased by the electron-donating inductive effect of the alkyl groups.

For example.

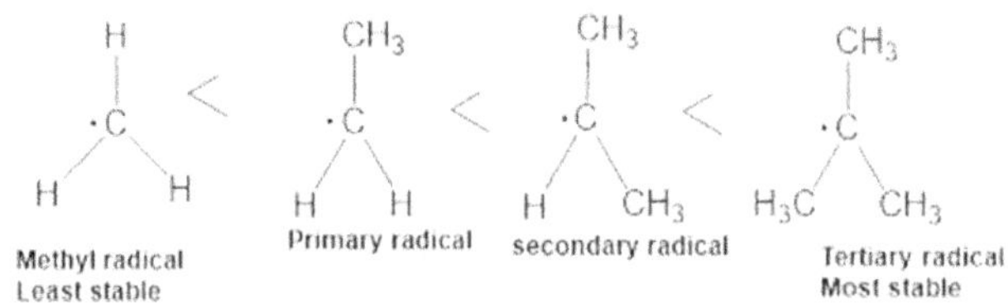

Similarly, electron withdrawing groups like -NO2, -CN, -COOH by withdrawing electron from radical center, reduces the electronegativity and electron density thereby decreasing the stability of free radicals.

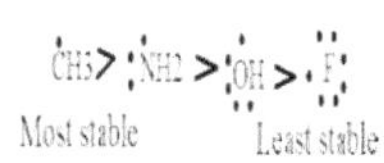

Resonance Effect. Stability of free radicals significantly depends on the delocalization of its unpaired electron through resonance. More the number of resonance structures greater delocalization and greater the stability.

Resonance effects cause benzylic and allylic radicals to be more stable than alkyl radicals because they allow an unpaired electron to delocalise over a conjugated pi bond system.

Table 1.

Ally radical	2 resonance structures	More stable
Benzyl radical	5 resonance structures	More stable
Alkyl radical	No resonance structure	Less stable

Hyperconjugation

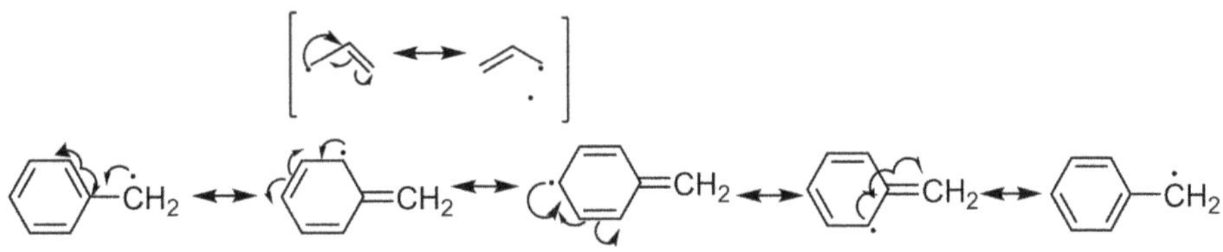

Increase in number of α – hydrogens increase the hyperconjugation effect, which in turn increases the free radical stability.

Note: Hyperconjugation cannot stabilise a carbon free radical as much as a carbocation. The reason is, the repulsion between the electrons in a carbon free radical is greater than that of a carbocation. In carbocation, the σ-bond orbital interacts with a neighbouring empty anti-bonding orbital when there is hyperconjugation. Hence carbocation is stabilised by the interaction between a filled and an empty orbital. But in the case of free radical at least one electron would be in the anti-bonding (or non-bonding) orbital.

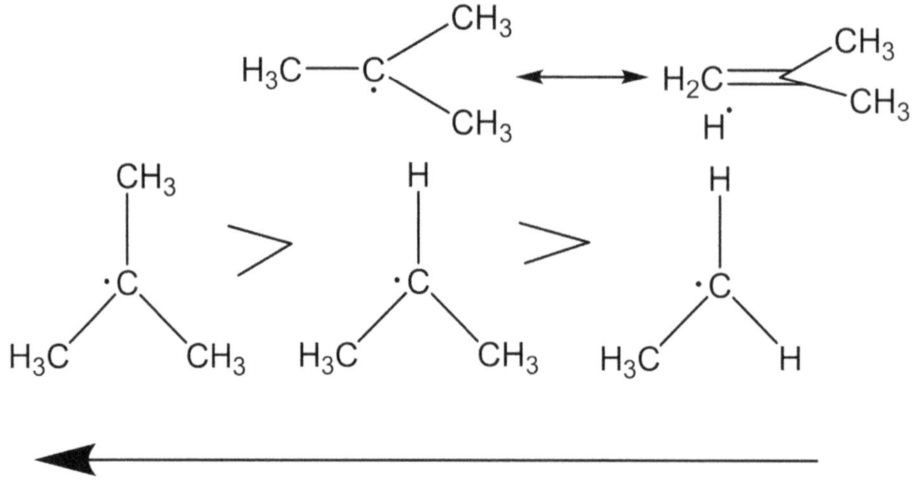

Stability increase

Hybridization. Hybridization also plays a major role in deciding the stability of free radical. sp hybridized orbital with 50% s and 50% p character decreases the stability of the free radical this is due to the close interaction of half-filled empty orbital with the nucleus. Hence as the s character increases the free radical stability decreases as shown below.

$sp^3 \longleftrightarrow sp^2 \longleftrightarrow sp$.

Adjacent Atoms with Lone Pairs. Free radicals are stabilised by atoms such as oxygen and nitrogen which donate its lone pair of electrons to the half-filled empty p orbital. Lone pair adjacent plays a considerable role in stabilising the free radicals. When comparing

$$\dot{C}H_3 \quad H_3C-\dot{C}H_2 \quad H_3CO-\dot{C}H_2$$

radical the later is more stabilised due to adjacent lonepair of electrons from oxygen atom.

a) Electronegativity of adjacent atoms

The presence of more electronegative group adjacent to the radical carbon higher is the bond strength and lesser its stability

b) Size of the atom

The size of free radical has a greater impact on its stability. Generally, a larger free radical tends to be more stable than smaller ones. For eg., as one moves down a column of the periodic table, the stability of free radicals increases: [$F^\bullet$ < $Cl^\bullet$ < $Br^\bullet$ < $I^\bullet$]. Because a larger amount of the electron-deficient orbital is exposed. Free radicals become less stable when electronegativity increases along the periodic table. Because the iodide radical's orbital is bigger and consequently spread out a larger volume, it is the most stable radical. A larger atom would feel a greater effective nuclear charge due to its closer proximity to the nucleus than a smaller atom.

Table 2.

Small size radicals	**Methyl, ethyl radicals**	**Highly reactive**	**Short lived**
Medium size radicals	Propyl, butyl radicals	moderately reactive	Relatively short lived
Large size radicals	Hexyl, octyl radicals	Less reactive	Relatively long lived
Macro size radicals	Polymer radicals	Highly stable	Long lived

Generation of Free Radical

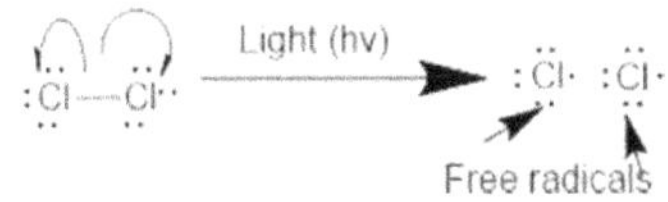

Free radical reactions do not entail the giving or receiving of an electron. Rather, it involves in homolytic cleavage of breaking of bonds so that an equal number of electrons is distributed to each atom.

Formation of Free Radicals generally occurs by the following three methods

By Thermal Reaction

At moderate temperature, compounds which have inrinsically weak bond (Dialkyl peroxides) and strong bond (AIBN) both undergo homolytic cleavage to form free radicals.

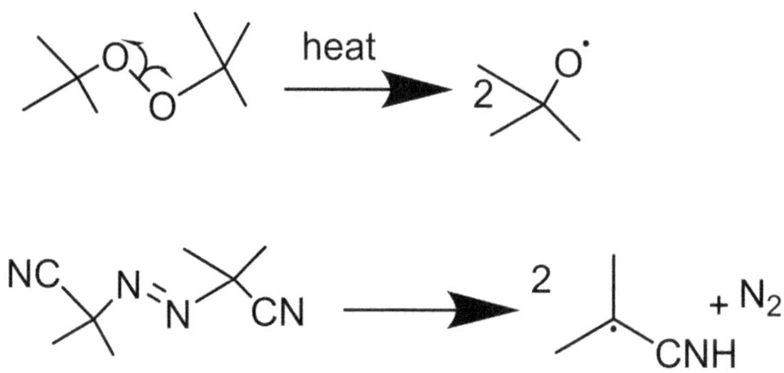

By Photochemical Reaction

When the energy greater than that of the bond (alkoxy radicals from alkyl nitrite or hypo chlorites) to be cleaved is irradiated with photochemical light, molecule undergo electronic excitation and homolytic cleavage occurs to produce free radical.

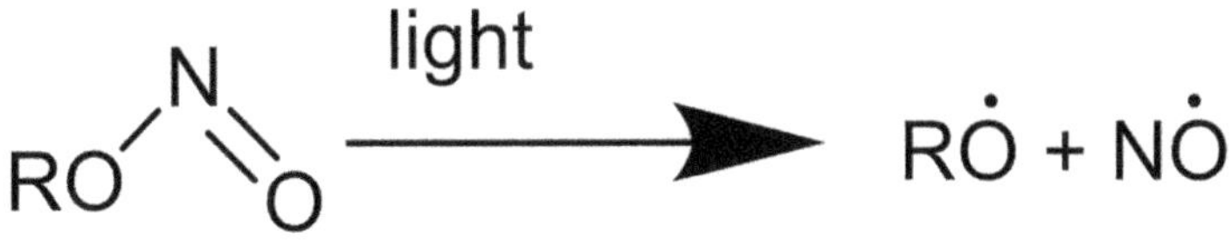

By Redox Reaction

Homolytic cleavage of covalent bond by transfer of electron. Through the process of electron transfer, covalent bonds can be disrupted by either accepting an electron from a donor or giving an electron to an acceptor.

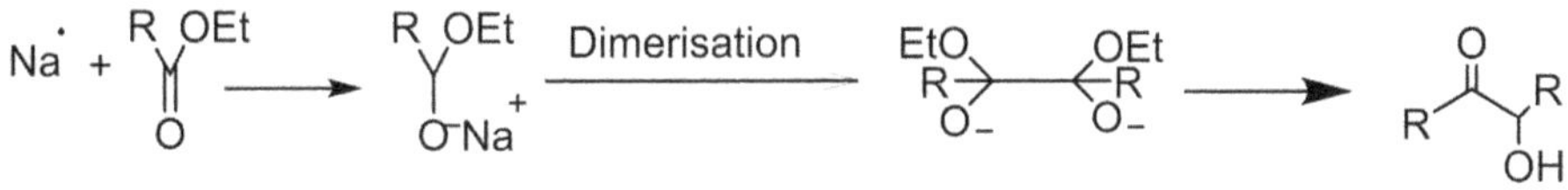

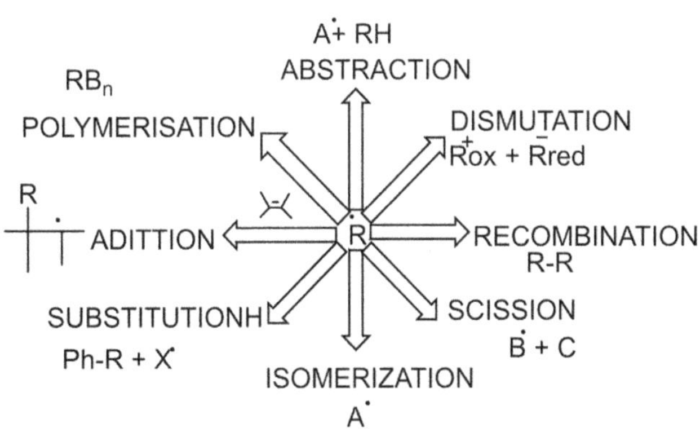

REACTIONS OF FREE RADICALS

Free Radicals Undergo The Following Reactions Mentioned Below.

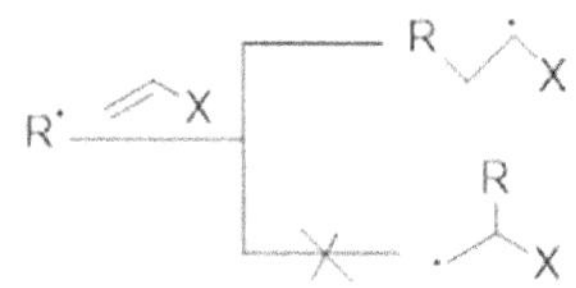

Abstraction

Free radicals abstract an atom from carbon to initiate reactions with saturated organic compounds. Bond dissociation energy and polar effects control the free radicals' selection for particular kinds of C-H bonds. As the energy of bond dissociation drops, the rate of abstraction process increases.

Addition Reaction

Free radicals undergo addition to carbon-carbon double bonds. The reaction is generally selective. For example, addition to CH2=CHX takes place exclusively at the methyl group, irrespective of the nature of X.

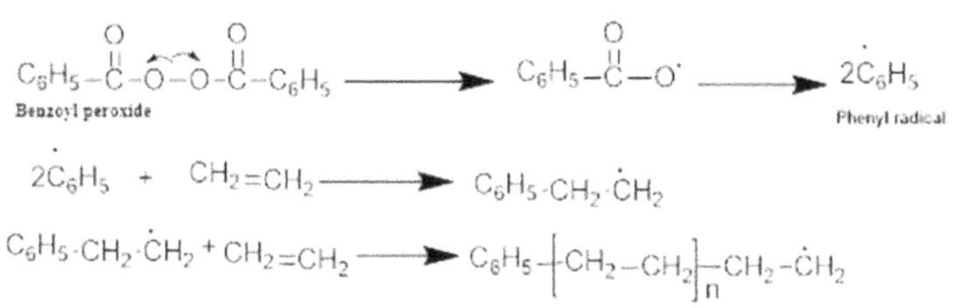

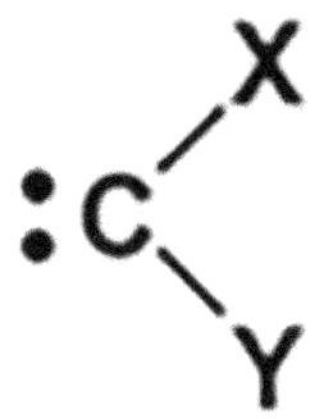

The first stage of the polymerisation process is called initiation. In order to start the polymerisation process, an active centre is created.

In the first step, the starting molecules yield one or two radicals.

In the second step, radicals are moved from the starting molecules to the monomer units. Once the free radical has begun to form, it targets a different monomer subunit. It forms a stable bond with another carbon atom by using one of the pi bond electrons. The molecule becomes a radical as a whole when the other electron goes back to the second carbon atom.

The chain continues to grow after initiation until either termination occurs or there is no more monomer.

CARBENE

Carbenes are electron deficient compounds with neutral divalent carbon atom (uncharged) surrounded by sextet of electron (Abramovitch, 2011).

The general structure of carbenes is

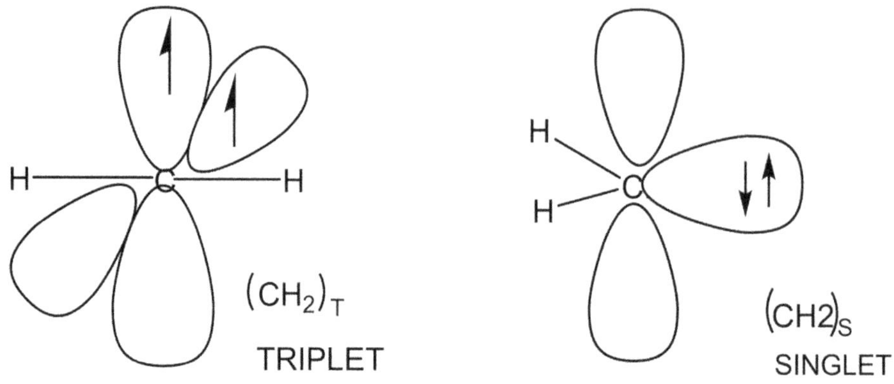

Structure

Carbenes are extremely reactive in nature, with lifespan nearly less than one second.

Carbenes have only been isolated. Exceptionally some can be trapped in matrices at low temperatures (77 K or less).

Divalent carbon compounds known as carbenes have two non-bonding electrons on a single carbon atom (a total of six electrons on carbon).

There are two significant options:

1. Singlet carbene
2. Triplet carbene

The formation of singlet or triplet is based on whether the two nonbonded electrons are paired or unpaired. The molecular electronic states decides whether it is a singlet state and triplet state. If the two nonbonded electrons are paired with antiparallel spin it is a singlet carbene with sp^2 hybridised carbon. If the two nonbonded electrons are unpaired with parallel spin it is a triplet carbene with sp hybridised carbon.

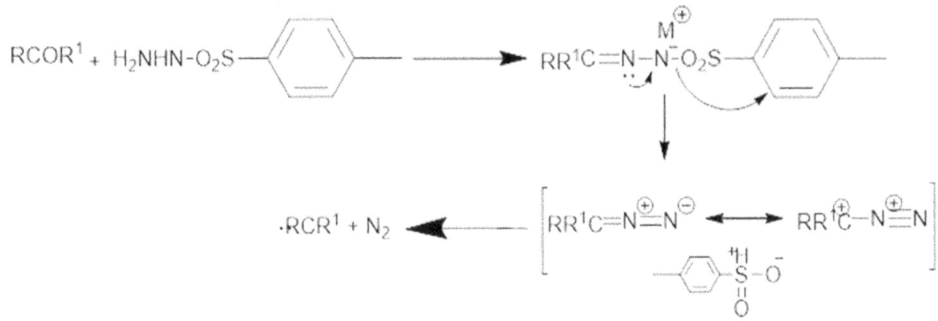

Therefore, one can say that singlet methylene should be diamagnetic and triplet methylene be paramagnetic a diradical.

Table 3.

	SINGLET CARBENE	TRIPLET CARBENE
ELECTRONS	Paired	Unpaired
GEOMETRY	Bent	Linear, Bent
HYBRID STRUCTURE	Sp^2	Linear (sp), Bent (sp^2)
MAGNETIC PROPERTIES	Diamagnetic	Paramagnetic

Generation of Carbenes

a) Preparation of carbenes from carbonyl compounds
b) Thermolysis or pyrolysis of Tri halomethyl mercury compounds
c) Trichloromethyl trihalosilane
d) Pyrolysis or Photolysis of ketenes, epoxides, aziridines

Preparation of Carbenes from Carbonyl Compounds

Several methods are available for synthesis of carbenes from carbonyl compounds. An aldehyde or a ketone is converted to a p-toluene sulphonyl hydrazone (Tosylhydrazone) which in turn is treated with a 1 eq. of base at low temperature ($NaOCH_3$) to form a salt of hydrazine followed by thermal or photolytic decomposition to form carbene.

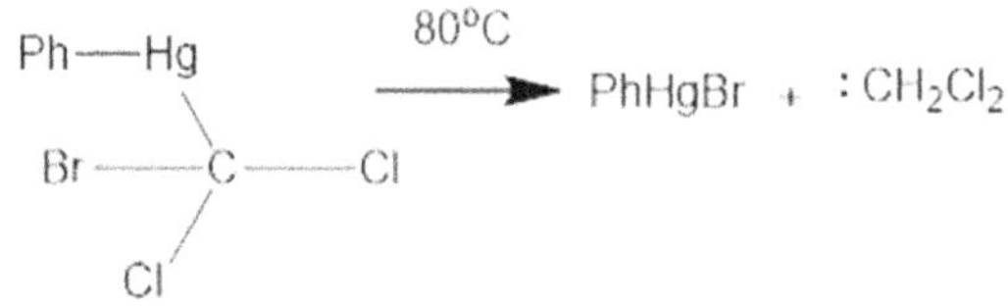

Thermolysis or Pyrolysis of Tri Halomethyl Mercury Compounds

Pyrolysis of Tri halomethyl mercury compounds occur at high temperature leading to break down of complex molecules into simple products like Hg, halogenated gases, and carbon containing products. Specific product and their yields depend on various reaction conditions like temperature, pyrolysis condition, structure etc.

Trichloromethyl Trihalosilane. *Formation of carbenes in near neutral media.*

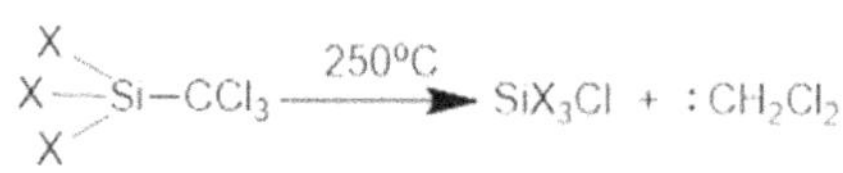

Pyrolysis or Photolysis of Ketenes, Epoxides, Aziridines

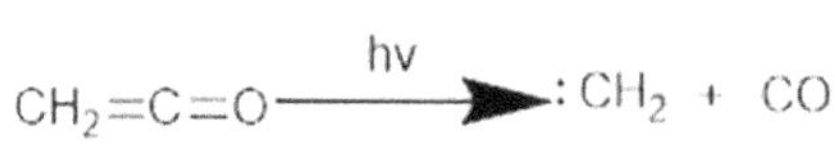

Ketene generate carbenes on photolysis or thermolysis as per the following reactions.

The epoxides on photolysis lead to the formation of carbenes.

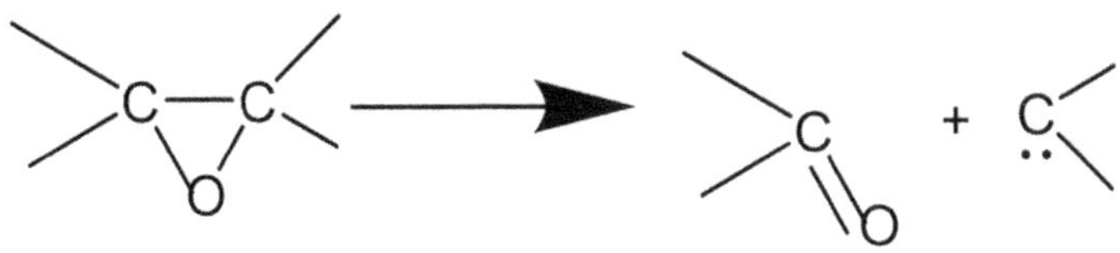

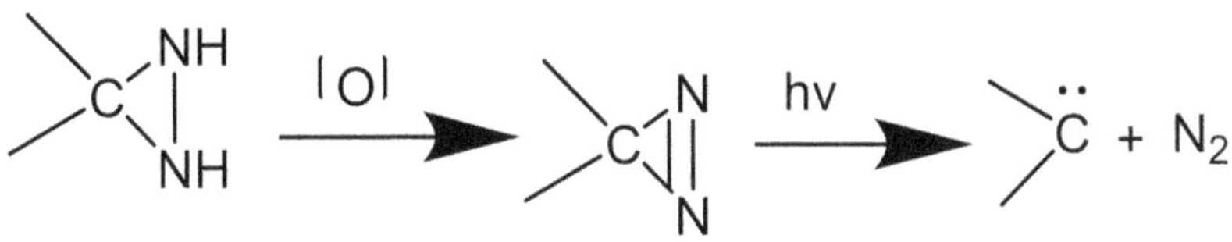

Diaziridines

Diaziridines three membered rings with two nitrogen can decompose photolytically to form carbenes

Reactions of Carbenes

Reimer-Tiemann Reaction

A phenolic aldehyde is prepared when a carbonyl compound is treated with chloroform in the presence of sodium hydroxide. This process introduces an aldehyde group (-CHO) at the ortho position of the benzene ring. The carbene formed through the deprotonation of the chloroform, which attacks the carbonyl carbon to give aldehydes.

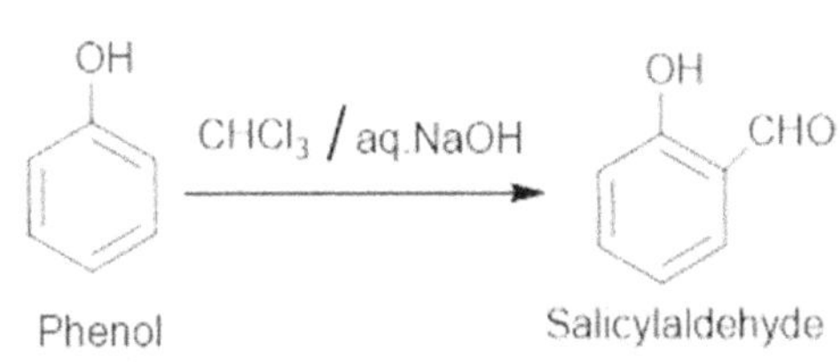

Insertion into C – H Bonds and Other Bonds. It has been noted that insertion reactions between alkanes and the activated carbenes produced by photolysis of diazomethane or ketene occur as shown below.

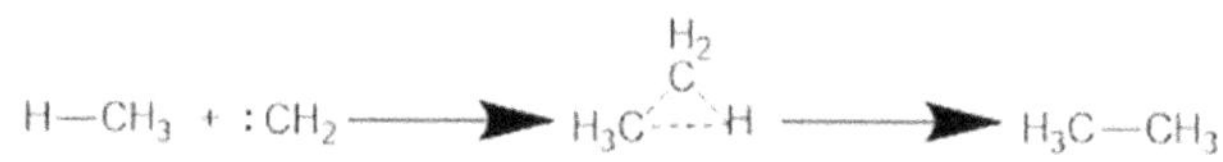

Wittig Rearrangement Reaction

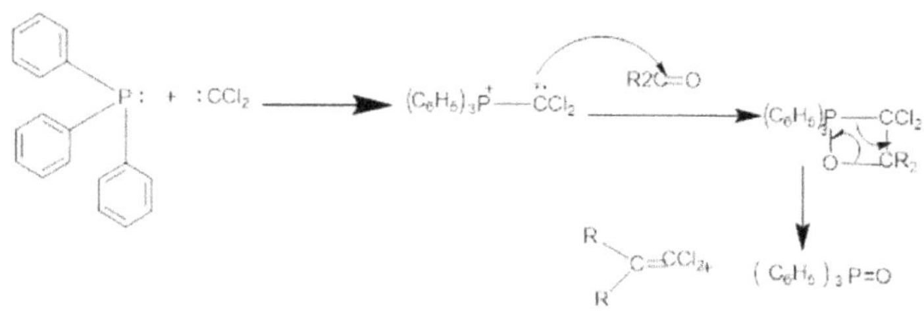

When dichlorocarbene and triphenylphosphine combine, a ylide intermediate is produced which on treatment with ketone 1,1-dichloroalkenes.

The Wittig reaction's mechanism is as follows

OVERALL SUMMARY OF REACTIVE INTERMEDIATES

Table 4.

Reactive intermediate	Hybridization	Shape	Structure
Carbocation	sp^2 hybridization	Trigonal planar	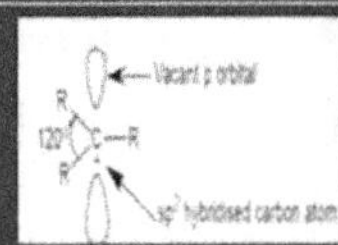
Carbanion	sp^3 hybridization	Pyramidal	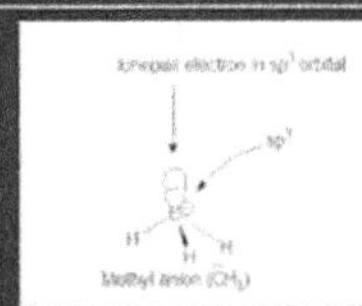
Free radical	sp^3 hybridization	shallow pyramidal geometry	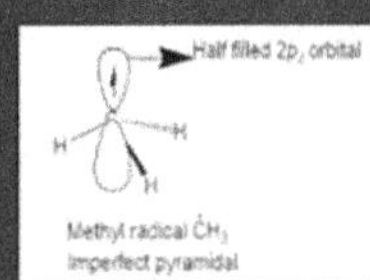

REFERENCES

Abramovitch, R. (2011). *Reactive intermediates*. Springer.

Ahluwalia,V. K., &Parashar,R. K.(2011). Organic reaction mechanisms.

Hornback, J. M. (2005). *Organic chemistry*. Cengage Learning.

Norman, R. O., & Coxon, J. M. (1993). *Principles of Organic Synthesis* (3rd ed.). CRC Press. DOI: 10.1007/978-94-011-2166-8

Santos, L. S. (2010). *Reactive intermediates: MS Investigations in Solution*. Wiley-VCH.

Singh, J., & Yadav, L. D. S. (2010). *Organic chemistry.* Revised Edition. Pragati Prakashan Publisher.

Smith, M. B., & March, J. (2006). *March's Advanced Organic Chemistry* (4th ed.). John Wiley & Sons, Inc. DOI: 10.1002/0470084960

Vajira. P.B. (2013). *Reactive intermediates.*

Chapter 5
Spectroscopic Methods for Studying Reaction Mechanisms

Charuwan Thanawiroon
https://orcid.org/0009-0004-9076-3219
Ubon Ratchathani University, Thailand

Bancha Yingngam
https://orcid.org/0000-0001-7215-9123
Ubon Ratchathani University, Thailand

ABSTRACT

Spectroscopic methods represent key tools for identifying organic reaction mechanisms and obtaining information about molecular structures and transient intermediates. However, the broad repertoire of organic reactions, frequent difficulties in tracking compound intermediates, and challenges in the interpretation of spectral data can hamper progress and lead to misconceptions. This chapter aims to explain these methods alongside kinetic studies and isotope effects to comprehensively understand reaction dynamics. This chapter explains how nuclear magnetic resonance, infrared spectroscopy, ultraviolet–visible spectroscopy, and computer-based mass spectrometry reveal interactions and highlights the importance of such approaches through comparison and spectral simulations. The finalized section demonstrates the need for a variety of perspectives and overall

DOI: 10.4018/979-8-3693-6473-4.ch005

explanations for promoting the synthesis of complex organic compounds. The field looks forward to significant progress with further real-time perspectives and education on reliable machine learning using spectroscopy.

INTRODUCTION

Returning to the question of the importance of studying reaction mechanisms, it is worth mentioning that understanding reaction mechanisms in organic chemistry is key to determining the basic processes that underlie chemical transformations (Anderson-Wile et al., 2024). This knowledge enables the determination of how structurally identical reactants can eventually turn into products, and already using this knowledge, chemists will be able to predict reactivity and selectivity trends to synthesize new pharmaceuticals, agrochemicals, and materials sciences (Koothradan et al., 2024). In addition, knowledge of the mechanisms of drug formation and the optimal reaction conditions will eliminate unwanted byproducts and only increase the yield of desired products (Bromfield Lee & Nelson, 2024). Moreover, the study of reaction mechanisms will further develop chemical theory and facilitate learning for future chemists, providing a thought focus to logically approach chemical problems (Venetos et al., 2024).

Among the most powerful modern tools for researching organic reaction mechanisms are spectroscopic methods. With their help, one can observe the dynamics and structure of molecules and their environments in real time under various conditions. The mainstream spectroscopy methods are as follows:

- Nuclear magnetic resonance (NMR) spectroscopy: This type of spectroscopy provides information about the molecule's backbone and is used to determine the atom's connectivity and molecular geometry (Anderson-Wile et al., 2024). NMR spectroscopy enables real-time monitoring of chemical reactions by tracking changes in the chemical shifts and peak intensities of reactants, intermediates, and products. This technique offers valuable insights into reaction kinetics and mechanisms.

- Infrared (IR) spectroscopy: IR spectroscopy is favorable for determining the presence of functional groups and studying the vibrations of molecules and can yield information on bond types (Hoenig, 2024). IR can be used to monitor the progress of chemical reactions by observing changes in the IR absorption bands of reactants, intermediates, and products. This allows for real-time tracking of the formation and consumption of different functional groups.
- UV–Vis spectroscopy: UV–Vis spectroscopy is predominantly used to analyze the electronic transition of molecules, i.e., their conjugation levels and electronic environments (Hoenig, 2024). UV–Vis spectroscopy enables the monitoring of chemical reactions by detecting variations in the absorbance of reactants, intermediates, and products at particular wavelengths. This method facilitates real-time tracking of concentration changes. The reaction rates and kinetic parameters can be determined by measuring the absorbance of specific species over time.
- Mass spectrometry (MS): MS is used mainly to determine the molecular masses of products, but with the development of tandem MS/MS techniques, the analysis of structural fragments has become possible (Hoenig, 2024). MS is a powerful tool for monitoring and determining the progress of chemical reactions. This approach provides detailed information about the reactants, intermediates, and products, providing insights into reaction mechanisms, kinetics, and pathways.

These findings are not exhaustive, and to understand the reaction mechanism in depth, several of these methods need to be used simultaneously.

The purpose of this chapter is to provide a comprehensive overview of the utility of various spectroscopic techniques that enable them to reveal organic reaction mechanisms. In each section, the readers will learn the basic principles on which each method operates, examples of its applications within kinetic studies and isotope effects and how combining spectroscopy with computational tools enables chemists to obtain a fuller picture of reaction dynamics. This chapter will further demonstrate examples from the relevant literature where these spectroscopic analyses have produced meaningful results in various complex organic reactions. By giving the readers access to cases of these techniques or the latest research, the authors hope to enhance the readers' comprehension. Additionally, this chapter will succinctly present the future developments of spectroscopic studies in organic chemistry. This will take into account the new applications of machine learning and real-time monitoring systems that promise to be revolutionary in this field. This chapter will cater not only to academics but also to practical readers who want to implement state-of-the-art methodologies in their industry.

SPECTROSCOPY FUNDAMENTALS

Principles of Spectroscopy

The science behind spectroscopy is based on the simple fact that light, or more generally, electromagnetic radiation, can be involved in interactions with matter. During this interaction, the latter may absorb, emit, or scatter light, thus doing so at particular wavelengths. Each molecule exhibits quantized energy states, and for the transitions between these states to occur, they must absorb or emit an amount of energy that corresponds precisely to the exact difference in these states. This difference in energy is localized to certain areas of the electromagnetic spectrum, which is why spectroscopy formally enables probing and investigating particular regions, and light at various wavelengths interacts with matter. Upon absorption of light, the molecule's electrons are excited to higher energy states that are again dependent on the molecule's structure, bonding, and chemical environment, meaning that the energy required for transitions is unique. As these excited electrons relax to a lower state, they emit energy in various other forms that can be systematically measured (Linne, 2024). Various spectroscopy methods measure this absorbed or emitted energy and display it as a spectrum with one axis showing the interaction's intensity and the other plotting the wavelength or frequency of the light. This spectrum forms a chemical substance's spectral fingerprint, as each chemical element or molecule extracts or emits unique interaction energy. These spectral fingerprints are essential for identifying chemical substances and their properties, interactions, and dynamics. Consequently, through this technique, spectroscopy is fundamental in chemistry, pharmacy, biology, physics, and materials science.

Types of Organic Chemistry Spectroscopies

Nuclear Magnetic Resonance

NMR spectroscopy is an important technique in organic chemistry. It is especially well known for its ability to yield information regarding the structure and dynamic composition of molecules. Chemists rely on NMR to reveal aspects of the molecular structure, dynamics, and chemical environment by taking advantage of the magnetic properties exhibited by certain atomic nuclei. This section elaborates on the concepts, principal methods, approaches, and usages of NMR in elucidating complex organic molecules.

Principles of NMR Spectroscopy. The concept of NMR is based on the magnetism of atomic nuclei or nuclear spin. Certain atomic nuclei, such as ^{1}H (proton) and ^{13}C (carbon-13), have an inherent property known as spin. This spin can create a magnetic moment, so those atomic nuclei behave as small bar magnets. Not all atomic nuclei are appreciable by NMR; only atomic nuclei with odd atomic numbers or odd mass numbers will emit the kind of magnetism necessary for NMR analysis. When placed in an exterior magnetic field, these magnetic atomic nuclei try to orient themselves through the magnetic field. Nevertheless, according to quantum mechanics, each atomic nucleus can have several quantized orientations about the guidance of an external magnetic field. If the atomic nucleus of interest has a spin number (I) of 1/2, then there will be only two orientation possibilities: alignment through the field, exploiting the lower energy state, or against the field, utilizing the high energy state. Thus, both ^{1}H and ^{13}C atomic nuclei have a spin number of 1/2 (Caceres-Cortes et al., 2024).

NMR spectroscopy, on the other hand, places the sample in a strong, consistent magnetic field established by a superconducting magnet. To inspect the nuclear environments in question, radiofrequency energy (RF) at a particular frequency equivalent to the energy disparity between the two spin states is employed. This frequency is often referred to as the Larmor frequency, which is distinctive to each type of nucleus and relies on the magnetic field intensity. When that frequency is utilized, nuclei in the lower energy level may absorb the RF energy and flip to the high-energy state; this action is called resonance. Consequently, the nuclear resonance energy needed for this move to reattain stability is incredibly delicate to the chemical surroundings of the nucleus. This is because the bodies of electrons all around the nucleus could hinder the shielding of the nucleus from the intensity of the external magnetic field. The most crucial feature of the use of NMR spectroscopy in identifying molecular architecture is chemical shifts. A chemical shift is the variation in a nucleus's resonance frequency produced by the nucleus's electronic environment. A chemical shift is similar to a fingerprint of a molecule in that it can be utilized to further determine the spatial relationships of specific nuclei within a sample. This shift is due to the shielding of electrons throughout a nucleus from an external magnetic field being shielded to various extents depending on the sample's electronic structure. Consequently, each shape of the nucleus in various chemical habitats vibrates at various frequency levels. The plotted frequency, chemical shift, and signal intensity, corresponding to the number of nuclei vibrating at any of the specified frequencies, are used to construct the NMR spectrum. The resonance frequency of the tested sample nuclei composes the peak of the NMR spectrum. Finally, the peaks' frequency, intensity, and multiplicity provide comprehensive molecular structure arrangements, including mastering particular atoms' frequency and relation as well as different unit groups' interconnection audit trail (Koothradan et al., 2024).

Instrumentation and Technique. NMR spectroscopy typically involves several key components (Figure 1):

- Magnet: The resolution and sensitivity of the NMR spectrum increase with the field strength of the magnet, which is typically measured with a Tesla. High-field magnets allow for better peak separation in the spectrum.
- RF transmitter and receiver: RF pulses excite nuclei, and a transceiver receiver detects signals generated by nuclei after their relaxation to equilibrium. To the best of the authors' knowledge, the RF energy

used by modern NMR instruments is the power of the magnetic field of the spectrometer. The higher the frequency of the RF radiation is, the stronger the magnet needed. Thus, NMR spectrometers are often categorized by the RF frequency they use, and the field strength is another way of thinking about the same thing. For instance, a common 300 MHz NMR spectrometer used in many chemistry laboratories typically operates at a fixed magnetic field strength of 7.05 Tesla and a radio frequency of 300 MHz. A 500 MHz NMR (sometimes referred to as a "500 Mega") is generally operated by a magnet with an 11.74 Tesla field and requires an RF energy of approximately 500 MHz. For example, the widely available 300 MHz NMR spectrometer found in many chemistry laboratories operates at a 7.05 Tesla magnetic field and 300 MHz RF energy. With an RF energy of approximately 500 MHz and 500 MHz NMR, a higher field instrument operates at approximately 11.74 Tesla. The 600 MHz NMR instrument runs at approximately 14.1 Tesla, and the RF radiation is approximately 600 MHz. The field strengths of the most frequently used academic and industrial research magnets are 800 MHz (18.8 Tesla) and even 1 GHz (23.5 Tesla), which indicate RF energies of 800 MHz and 1 GHz, respectively.

- Sample preparation: The sample can be dissolved in solvents; many of those solvents are deuterated to minimize interference from ^{1}H in the solvent molecules. In NMR spectroscopy, a few of the common solvents that can be used to dissolve samples are deuterated chloroform ($CDCl_3$), a broadly used solvent as a universal solvent for NMR, which is a barely soluble organic compound whose spectrum is rather simple; deuterated dimethyl sulfoxide ($DMSO - d_6$), which has a high boiling point and can dissolve a wide range of polar compounds; deuterated methanol (CD_3OD), which is beneficial for polar substances and is especially useful in biological NMR spectroscopy; and deuterated acetone (CD_3COCD_3), which is employed in samples that may interact or react with alcohols. It is a good solvent for many organic compounds, and compared with simple esters, it is very inert, and deuterated water (D_2O) is useful for NMR studies in aqueous solutions, particularly for biochemical and enzymatic reactions.

- Probe: A device that contains a sample and fits into the magnet bore, producing a uniform RF field through the sample and detecting the NMR signal emitted by the sample.

Figure 1. Key components of an NMR spectroscopy setup

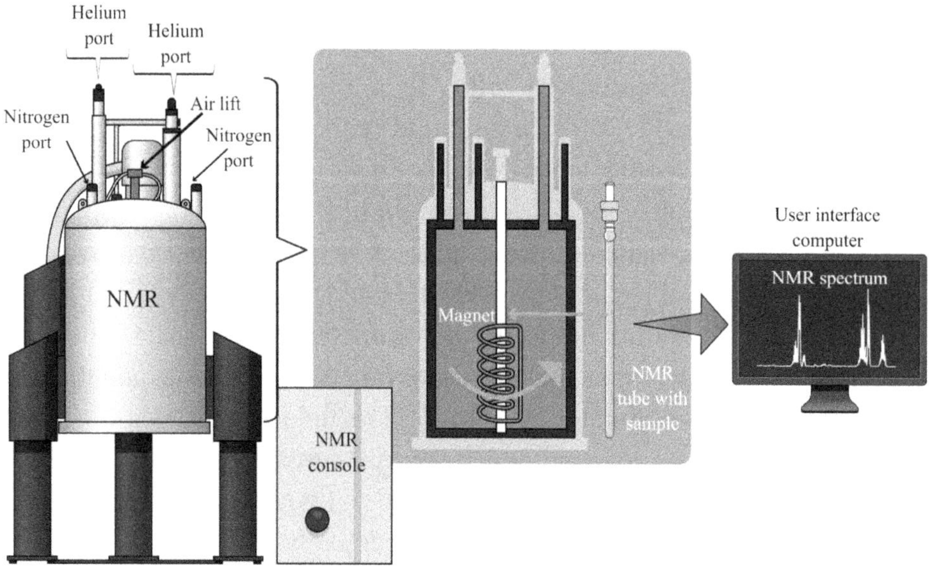

(Authors)

NMR Spectroscopy Type. The common types of NMR spectroscopy are as follows.

- Proton NMR (^{1}H NMR): This is the most common type of NMR spectroscopy. It is primarily utilized to identify the types and environments of hydrogen atoms in a molecule. Scientists can interpret structural details about the molecule by analyzing the absorption spectra produced when hydrogen nuclei in a sample are exposed to a strong magnetic field and a radiofrequency pulse.
- Carbon-13 NMR (^{13}C NMR): ^{13}C NMR focuses on the carbon atoms of organic compounds. This approach provides detailed information about the carbon skeleton of molecules. Despite its lower sensitivity than proton NMR due to the lower natural abundance of ^{13}C,

this type provides valuable information about the carbon skeleton of organic compounds. This technique can show how carbons are chemically shifted by neighboring groups and can be enhanced by techniques such as decoupling (removing interactions with hydrogen atoms) or dynamic nuclear polarization to improve sensitivity.

Table 1 compares ^{1}H and ^{13}C NMR spectroscopy for elucidating organic compounds. It is evident from the table that ^{1}H NMR is more sensitive, given that hydrogens are more abundant in the environment and are also more sensitive. It is also clear that this method produces a higher resolution spectrum, which is useful for highly structured determinations in the chemical shift range of 0–14 ppm. It is also useful in the determination of functional groups and thus provides extensive coupling information, which is always useful in structure determination. It is concluded that proton NMR is a more useful technique in general. In comparison, ^{13}C NMR is less sensitive given the low natural abundance of carbon and the insensitivity of carbons in general. Although it is more useful in the determination of the skeletal position of every carbon and is still useful in the determination of the structure of any molecule, it has lower resolution and requires more concentrated sample identification, although it generally requires longer acquisition times for quantitative analysis.

Table 1. Comparison of ^{1}H NMR and ^{13}C NMR spectroscopy in organic compound elucidation

Feature	Proton (^{1}H) NMR	Carbon (^{13}C) NMR
Nuclear spin	1/2	1/2
Natural abundance	99.98%	1.1%
Sensitivity	High (most sensitive NMR active nucleus)	Lower compared to ^{1}H NMR due to lesser abundance and lower gyromagnetic ratio
Resolution	High resolution with complex but informative spectra	Lower resolution but simpler spectra due to fewer carbons than hydrogens

continued on following page

Table 1. Continued

Feature	Proton (^1H) NMR	Carbon (^{13}C) NMR
Chemical shift range	0–14 ppm	0–220 ppm
Typical applications	Structure elucidation of organic compounds is especially useful for identifying functional groups	Important for determining the framework of carbon skeletons in organic molecules
Quantitative analysis	Easier and often used for quantitative analysis	More challenging, often requiring lower acquisition times
Coupling information	Provides extensive coupling information (*J*-couplings) between hydrogen atoms	Coupling usually observed between carbon atoms and directly bonded protons
Sample preparation	Generally, requires less sample; can use dilute solutions	Typically, requires more concentrated samples due to lower sensitivity

To clarify these results, Table 2 provides a detailed comparison of the ^1H and ^{13}C NMR chemical shifts for the functional groups, along with detailed information on their molecular environments. The ^1H and ^{13}C NMR signals of the methyl groups shifted from 0.8 to 1.2 ppm and 10 to 20 ppm, respectively, which is characteristic of the groups attached to saturated carbons. Methylene and methine groups present shifts that demonstrate their involvement in saturated hydrocarbons and cyclic structures, specifically, the second type of group in methylene indication. Alkenes and aromatic rings show greater downfield shifts for alkenes resulting from the deshielding impact of the pi bonding: 4.6–6.0 in ^1H and 110–150 in ^{13}C NMR, and aromatic stabilization: 6.5–8.5 and 110–160 ppm for aromatics, while the others include significant shifts toward downfield resulting from the strong electronegative influence of the adjacent oxygen atoms. The range of shifts of the esters, alcohols, and amines demonstrates that the shifts of the functional groups differ within compounds, specifically depending on the location of bonding and hydrogen bonding of the functional group surroundings, which is critical for understanding the chemical properties and interactions of these groups within molecules.

Table 2. ^{1}H NMR and ^{13}C NMR chemical shifts and corresponding molecular structures

Functional Group	[1]*H* NMR	[13] NMR	Description of Molecular Environment
Methyl group (CH_3)	0.8–1.2	10–20	Shifts for methyl groups, generally attached to saturated carbons.
Methylene group (CH_2)	1.2–1.8	20–40	Shifts for methylene groups in saturated hydrocarbon chains.
Methine group (CH)	1.4–2.6	25–50	Shifts for methine carbons are often indicative of CH groups in cyclic structures.
Alkene (CH_2=CH–)	4.6–6.0	110–150	Shifts associated with protons and carbons in alkenes.
Aromatic (Ar–H)	6.5–8.5	110–160	Shifts characteristic of aromatic hydrogens and carbons.
Aldehyde (–CHO)	9.0–10.0	190–200	High downfield shifts for hydrogen and carbon in aldehydes.
Carboxylic acid (–COOH)	10.0–12.0	170–182	Very downfield shifts for hydrogen and carbon in carboxylic acids.
Ester (COO)	3.5–4.5	160–185	Shifts for ester functionalities, reflecting ester linkage environments.
Alcohol (–OH)	1.0–5.5	50–80	Broad range due to hydrogen bonding, typically broad peaks for alcohols.
Amine (–NH_2)	1.0–5.0	30–50	Shifts for amines can vary widely depending on the environment and degree of hydrogen bonding.

Note: The values provided are indicative and can vary based on molecular context, such as electronic effects from nearly functional groups or the solvent used during NMR analysis.

- Two-dimensional (2D) NMR: 2D NMR techniques are essential for elucidating the structures of complex molecules for which one-dimensional techniques are insufficient. Techniques such as COSY (correlation spectroscopy), HSQC (heteronuclear single quantum coherence), and NOESY (nuclear Overhauser effect spectroscopy) (Table 3) provide information on how atoms within a molecule are connected to each other. For example, COSY reveals scalar coupling between hydrogen atoms that are typically separated by up to three bonds. This helps to establish connectivity and confirm molecular structures (Osifová et al., 2024). For HSQC, this technique provides a correlation between hydrogen and directly bonded carbon (or other heteroatom) nuclei, which is particularly useful in organic and biomolecular chemistry for mapping carbon–hydrogen frameworks in molecules (Caceres-Cortes et al., 2024). In addition, NOESY is useful for identifying the spatial proximity of atoms within a molecule. NOESY can show interactions between nonbonded atoms that are close in space. This provides crucial insights into the three-dimensional conformation of the molecule (Sancéau et al., 2024).

Table 3. Comparison of NMR spectroscopy techniques involving COSY, HSQC, and NOESY

Feature	COSY (Correlation Spectroscopy)	HSQC (Heteronuclear Single Quantum Coherence)	NOESY (Nuclear Overhauser Effect Spectroscopy)
Technique type	Homonuclear 2D NMR	Heteronuclear 2D NMR	Homonuclear 2D NMR
Main use	Determination of scalar couplings between protons that are through-bond.	Correlation of directly bonded hydrogen and carbon/ nitrogen atoms.	Identification of spatial proximity (less than 5 Å) between nonbonded protons.
Information provided	Reveals how atoms within a molecule are connected by chemical bonds.	Links hydrogen atoms with their directly bonded carbon or nitrogen atoms, crucial for determining backbone structures of organic compounds.	Provides insights into the three-dimensional structure and dynamics of the molecule by observing interactions between spatially close protons.
Sample requirements	Requires less sample concentration compared to NOESY.	Generally, requires isotopic labeling for detecting ^{13}C or ^{15}N, useful for larger biological molecules.	Requires higher concentrations to enhance the weaker NOESY signals.

continued on following page

Table 3. Continued

Feature	COSY (Correlation Spectroscopy)	HSQC (Heteronuclear Single Quantum Coherence)	NOESY (Nuclear Overhauser Effect Spectroscopy)
Complexity of spectra	Relatively simple spectra show cross-peaks only between coupled protons.	More complex due to its heteronuclear nature but provides clear correlations between different atom types.	Complex due to the inclusion of all interactions within 5 Å, including through-space coupling.
Typical applications	Used extensively in small molecule NMR for structure elucidation.	Predominantly used in protein and nucleic acid NMR to establish connectivity.	Essential in studying the conformations and interactions in biomolecules like proteins and nucleic acids.

Applications in Organic Chemistry. NMR spectroscopy plays a key role in organic chemistry, and its impact is evident in many application areas. First, NMR, which aims to determine the structure of organic compounds, is widely used in structural elucidation. This application area includes the identification of functional groups and the determination of the structure of the carbon–hydrogen framework. Another area is conformational analysis, where the technique tries to determine the different arrangements of the atoms in a given molecule and offers tremendous help in understanding molecular conformation dynamics. However, the use of NMR in chemical exchange analysis is important. NMR techniques have been used to study the amount of exchange between atoms or groups in a molecule and help scientists predict the reactivity and chemical characteristics of the molecule of interest. In addition, quantitative NMR also plays a significant role in organic chemistry due to the benefits of using NMR signals to obtain accurate elemental content in a mixture. This technique is applicable for assessing the concentration of mixture components, which renders it important for both qualitative and quantitative chemistry (Ma et al., 2024).

Infrared Spectroscopy

IR spectroscopy is an analytical method used in organic chemistry to identify and determine molecular structures based on vibrations. The method uses the interaction of IR radiation with the vibrations of molecules; its purpose is defined by the variance of the molecular dipole moment. In this section, the concepts and applications of IR spectroscopy, which is applicable in organic chemistry, are discussed.

Principles of IR Spectroscopy. IR spectroscopy is based on the concept that molecules absorb individual IR light at different frequencies, which consequently causes various bonds in the molecule to vibrate. Thai is the stretching, bending, twisting, and scissoring of the bonds between atoms. This energy absorbed from the IR light promotes these molecules from the ground vibrational state to an excited or higher vibrational state. The frequency at which a particular chemical bond absorbs IR light depends mainly on the following two factors. (1) As the mass of the atoms increases, the heavier atoms move slowly, meaning that the bonds including ponderous atoms will vibrate at lower frequencies than those among lighter atoms. (2) The strength of the bond: The double bond, or bonds including severely electronegative atoms, will also vibrate at high frequencies. When the bond strength decreases, such as for bonds between two similar atoms or single bonds, the vibrational frequency decreases. A bond of every kind exhibits a specific absorption spectrum for IR radiation, which can be a qualitative marker for distinctive functional groups within the molecule. When IR radiation is allowed to pass through a sample, certain frequencies of the radiation are absorbed by matching the natural vibration frequencies of the bonds within the molecules of the unknown samples. The absorption of energy results from the increase in the amplitude of the vibrational motion of those bonds. In terms of the IR spectrum, the analysis results for IR spectroscopy are usually displayed as a spectrum with a y-axis plot of the intensity of absorption on the y-axis versus frequency or a wavenumber x-axis plot of the frequency (commonly reported in wavenumbers, unit: inverse centimeters). Generally, the peaks in the IR absorption spectra are found at the frequencies at which a sample absorbs IR radiation. The positions or wavenumbers of these peaks depict the exact vibrations of bond motions, while the intensity of the absorption offers clues about the amount of absorbing material (Jirasek et al., 2024).

Instrumentation and Technique. The key components of an IR spectrometer include the following (Figure 2):

- Source: The effective source for an IR spectrometer is a blackbody radiator. This type of radiator can perfectly emit a continuous array of infrared light. Additionally, the blackbody source is named so that it can absorb all the radiation incident from electromagnetic radiation, independent of the frequency and the radiation angle, and is good for radiation. In an IR spectrometer, the radiation emitted covers the infrared wavelengths necessary for the excitation of diverse molecular vibrations in the sample.
- Monochromator: The monochromator is a vital part employed in an IR spectrometer to measure discrete wavelengths of radiation irradiating from the source. The reason is that various molecular bonds decay IR radiation at different wavelengths. The monochromator disperses the infrared light by employing a diffraction grating or a prism. Afterward, the device allows the desired wavelength of the spectrum to pass through a hole for sampling, enabling the analysis of molecular vibrations at the desired interval.
- Sample holder: The sample holder, termed the sample cell, should be suitable for hosting samples in any state. The level of design features differs with respect to the sample state; for example, liquids can be inserted between pieces of a high-IR transmissive object, for instance, potassium bromide. Moreover, gas might need a cell with an extended cell path to ensure competent interactions with IR radiation. The material used to form the sampling cell is vital to avoid absorption or interference at the targeted wavelengths.
- Detector: The detector in an IR spectrometer plays a crucial role in converting the infrared light that has passed through the sample into a measurable signal. Several types of detectors are employed based on attempt sensitivity and variation. Some of the general detectors employed include thermocouples, bolometers, and photovoltaic cells or photoconductive cells that are formed using materials such as mercury cadmium telluride. The detectors are responsive to

changes in light intensity and are effective at translating the changes to electrical signals that are processed digitally for analysis.

- Spectrometer: This is a system that records the intensity of the IR radiation after interaction with the sample. It then movies the data against wavelength or wavenumber, resulting in peaks at the wavelengths where the samples absorb the IR, showing different molecular vibrations in the sample. Usually, the spectrometer is implemented in a computer system that has software that allows data acquisition, analysis, and display. Thus, through stretching and vibration, scientists can interpret spectral data to identify functional bonds and molecular structures.

Figure 2. Instrumentation and technique of an infrared spectrometer

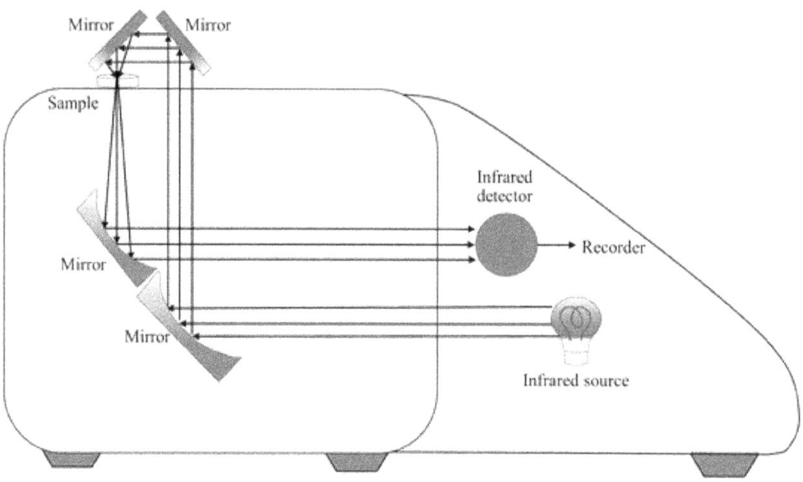

(Authors)

Types of IR Spectroscopy. IR spectroscopy is a fundamental analytical technique used across various scientific fields for identifying molecular structures and studying their interactions. Among the various methods of IR spectroscopy, Fourier transform infrared (FTIR) spectroscopy and attenuated total reflectance (ATR) spectroscopy stand out due to their specific applications, efficiency, and ease of use. The details of these two types of instruments are as follows:

FTIR Spectroscopy. FTIR spectroscopy (Table 4) achieves this efficiency by utilizing an interferometer mechanism, particularly the Michelson interferometer, which completely changes the way spectral data are acquired. Unlike dispersive instruments, which require moving physical optics to scan through different IR light frequencies, the FTIR instrument uses an interferometer to acquire all wavelengths simultaneously. The mechanism relies on splitting a single beam of IR light into two paths. One is reflected off a fixed mirror, and the other is reflected off a movable mirror. When the beams recombine, the waves interfere with each other. As the mirror moves, the interference pattern changes, from which the mirror position can be computed. This pattern, known as an interferogram, is recorded as the mirror shifts; hence, all of the IR frequencies are collected at once. The interferogram is transformed into a spectrum using Fourier transformation. Hence, FTIR can measure spectral data very rapidly; it only takes a few seconds since it acquires all of the information in the same instance. This is because fast data collection and FTIR are widely applied in research and industrial settings for different applications that usually necessitate detailed chemical information about organic compounds, polymers, pharmaceuticals, and even forensic samples. FTIR is particularly suitable for use in quality control, environmental monitoring, and clinical diagnostics where rapid identification is required due to its speed and outstanding resolution.

ATR Spectroscopy. One variant of IR spectroscopy, ATR spectroscopy (Table 4), is used to overcome some of these challenges related to sample preparation. In ATR, the IR light is aimed at a crystal with a high refractive index, and the sample is tested in direct contact with the exterior of the crystal. The IR light reflects internally at the interface between the crystal and the sample, causing only a small portion of the light, the evanescent wave, to be absorbed by the sample, usually only a few micrometers in depth. This allows for simplified sample preparation since the light does not need to pass through the sample. For this reason, ATR is typically used to analyze thick or impenetrable materials where thin sections cannot be produced. This is especially beneficial for solid or viscous samples. Additionally, the light penetration depth is dependent on the wavelength and angle of incidence of the light, as well as the refractive indices of the sample and the crystal, allowing surface analysis or scanning of thin films to be performed using ATR. Therefore, ATR is used in food science, pharmaceuticals, and chemistry and can be applied in suit for monitoring chemical reactions. The relative simplicity of sample preparation enables rapid and repeatable ATR measurements.

Table 4. Comparison of FTIR and ATR spectroscopy techniques

Feature	FTIR Spectroscopy	ATR Spectroscopy
Principle	Measures the absorption of infrared radiation by the sample, causing molecular vibrations.	Utilizes internal reflection and measures the evanescent wave absorbed by the sample.
Sample preparation	Often requires thin film or KBr pellet preparation.	Minimal preparation, samples can be analyzed in their natural state (solid, liquid, or gel).
Sensitivity	High sensitivity, especially with modern interferometers.	Generally, less sensitive than FTIR, but very effective for surface or thin-layer analysis.
Penetration depth	Measures bulk properties	Analyzes a thin layer on the surface of the sample (typically o.5 to 2 microns).
Applications	Suitable for quantitative and qualitative analysis of organic and inorganic compounds.	Ideal for coatings, films, and surface layers; often used when sample preparation is difficult.
Advantages	Provides precise measurement of the entire IR spectrum. Can analyze a wide range of samples.	Direct analysis with little to no sample preparation. Less risk of sample damage or contamination.
Limitations	Sample preparation can be complex and time-consuming.	Limited to surface analysis; depth information is not obtainable.

For IR spectroscopy, Table 5 provides the absorption bands and the corresponding molecular vibrations for various functional groups. These ranges are usually characterized by either a sharp peak or a medium-to-strong peak with sharp or broad peaks. Each of these may therefore represent different vibrational modes that are responsible for molecular bonding and molecular structure. It is evident that alkanes show the principal bands of sharp to medium per strong peak observed at 2860–2960 cm^{-1} due to C–H stretching. However, alkenes and aromatics are observed to exhibit C=C stretching; the only difference is that aromatic compounds are present at 1450–1650 cm^{-1}, while alkenes are present at highly variable wavenumbers, which indicates a difference in molecular structure. Moreover, other functional groups, such as aldehydes, ketones, esters, and carboxylic acids, also displayed strong to sharp bands at 1710–1750 cm^{-1}, which are attributed to C=O stretching. However, a slight difference is observed in the wavenumber range for each of the groups. In addition, carboxylic acid and alcohol functional groups show strong broad bands (2500–3300 cm^{-1}). Amine and amide functional groups display N–H interactions strongly influenced by hydrogen bonding, and their amides also show an additional peak at 1640–1680 cm^{-1}, indicating that C=O stretching is also responsible for the amide bands. Finally, a sharp peak at 2240–2270 cm^{-1}, which is also responsible for C≡N stretching, is observed for nitriles.

Table 5. IR absorption bands and corresponding molecular vibrations

Functional Group	**Wavenumber Range (cm^{-1})**	**Description of Vibrational Mode**
Alkanes (C–H stretch)	2850–2960	Sharp, medium to strong peaks for C–H stretching in alkanes.
Alkenes (C=C stretch)	1620–1680	Medium to strong peaks, indicative of C=C stretching in alkenes.
Aromatic (C=C stretch)	1400–1600	Multiple sharp peaks due to aromatic C=C stretching.
Aldehydes (C=O stretch)	1720–1740	Strong, sharp peak characteristic of aldehyde C=O stretching.
Ketones (C=O stretch)	1705–1725	Strong, sharp peak for C=O stretching in ketones.
Carboxylic acids (O–H stretch)	2500–3000	The broad, strong peaks due to O–H stretching, often overlapped.
Carboxylic acids (C=O stretch)	1700–1725	Strong, sharp peak characteristic of carboxylic acid C=O stretching.
Esters (C=O stretch)	1735–1750	Strong, sharp peak for ester C=O stretching.
Alcohols (O–H stretch)	3200–3600	The broad, strong peak is due to hydrogen-bonded O–H stretching.
Amines (N–H stretch)	3300–3500	Medium, sharp, or broad peaks for primary and secondary amines.

continued on following page

Table 5. Continued

Functional Group	Wavenumber Range (cm^{-1})	Description of Vibrational Mode
Nitriles (C≡N stretch)	2210–2260	Medium, sharp peak indicating C≡N stretching in nitriles.
Amides (N-H stretch)	3100–3500	Broad, medium intensity due to N–H stretching in amides.
Amides (C=O stretch)	1630–1690	Strong, sharp peak characteristic of amide C=O stretching.

Note: The wavenumber ranges provided can vary slightly depending on the molecular environment and should be interpreted in the context of the overall spectral data.

Applications in Organic Chemistry. IR spectroscopy has numerous uses in organic chemistry and is responsible for several instrumental functions. This approach is applicable because it allows for the identification of functional groups, which are characterized by the presence of specific absorption bands. For example, carbonyls, hydroxyls, and amines, among other groups, have such bands. Thus, chemists can compare them with known patterns to determine what functional groups are contained in a given compound. However, it is not the only function of IR spectroscopy, as it also contributes to molecular structure elucidation. This is because the data can provide information about the geometry and bond patterns that are essential for confirmation. Moreover, it is a tool for quantitative analysis; for example, the intensity of the IR band can be utilized to identify the concentration with the help of the Beer–Lambert Law. Finally, IR spectroscopy was used to investigate molecular interactions and dynamics. For instance, the alteration of the IR spectrum indicates the existence of hydrogen bonding, while changes can signal chemical reactions. Thus, it is a viable tool for understanding and manipulating organic compounds.

Ultraviolet–Visible Spectroscopy

UV–Vis spectroscopy is a basic but important analytical technique for studying molecular structure and dynamics. It is based on the focus of the absorption of a molecule by electrons, generally in the ground state and frequently in several excited states, as they are exposed to ultraviolet or visible light. Although termed UV/Vis, the same instruments can be utilized to include both spectra, there is no reason not to do so. This section will concentrate on the basis of this method, where it can be applied, and on which kind of molecular system one can obtain meaningful insight.

Principles of UV–Vis spectroscopy. The principle used in UV–Vis spectroscopy is the absorption of light by molecules. If the energy of the photon corresponds to the energy variation between two electronic states of a molecule, then the photon is subsequently absorbed by the molecule. This process results in electron promotion from low energy to a higher energy orbital. The key to this process is the molecule's electronic structure and which wavelengths it can absorb efficiently. UV–Vis spectroscopy is a nondestructive technique that is frequently used for the determination of light absorption in the potential power range of the electromagnetic spectrum, between 200 nm and 800 nm. A molecule absorbs photons of light, and the molecule's electrons are excited to higher-energy states (higher-energy orbitals). Common transitions observed via UV–Vis spectroscopy include π to π^* and n to π^*, which are evident in molecules with conjugated π systems and nonbonding electrons, respectively. This occurs for the n- π^* and π to π^* transitions in conjugated double bonds (pi bonds) for π to π^*. This means that the excitation process transforms the electrons in the bonding π orbitals into antibonding π^* orbitals. Conjugation can be seen particularly in organic compounds with long e-contiguous systems and causes downward energy transitions (UV absorbance in the near-UV region: wavelength of UV). When n transitions to π^*, the nonbonding orbital (n) is excited to become an antibonding π^* orbital. A select electron pair would most commonly be found in molecules with atoms such as oxygen or nitrogen; the transition from the energy required to excite an electron from the σ^*-state bond to the σ-state bond is generally easier to overcome than the energy required for the double bond in the π to π^* transition; thus, this type of transition is usually found in molecules that contain carbonyl moieties, that is, ketones and aldehydes. These transitions absorb light in characteristic energy (or wavelength) ranges that are specific to the light-absorbing molecule and can be used to gain detailed information about the presence and environment of given types of chemical bonds in the molecule as a whole. A UV–Vis spectrophotometer produces a spectrum of absorbance or transmittance versus wavelength. The spectral characteristics of the UV–Vis spectrum are as follows:

- Absorption peaks: These peaks are present where the substance absorbs the light. For the corresponding light wavelengths, each peak indicates the type of chemical bond, environment or electronic transition to which the peak corresponds.
- Wavelength: The position of the peak in terms of the peak's maximum is taken in terms of nanometers (nm), and the peak indirectly identifies the type of electronic transition and informs the reader about the functional groups present in the molecules.
- Absorbance: The intensity at which the absorption peak occurs if the sample is a strong or a weak absorber of light at a specific wavelength. It can also be beneficial when determining the concentrations of absorbing species in a sample on the basis of the Beer–Lambert law.

In analytic chemistry, Beer–Lambert's law is one of the simplest equations. The law stipulates that the absorbance of light passing through a material is proportional to the inherent properties of the passing material. The absorbance (A) is usually calculated as the material passing through the concentration (c) of the absorbing species in solution, the path length (l) of the sample, which is usually expressed in centimeters, and the molar absorptivity (ε), which is (Eq. 1):

$$A = \varepsilon c l \tag{1}$$

Table 6 outlines the typical UV–Vis spectroscopy results for various types of organic compounds, detailing their characteristic absorption wavelengths (λ_{max}), the electronic transitions involved, and a description of their spectral features. For instance, alkanes exhibit very weak absorption between 175–200 nm due to σ to σ^* transitions, reflecting the saturated nature of their hydrocarbons. Alkenes exhibit moderate absorption between 175–195 nm due to π to π^* transitions associated with their double bonds. Aromatic compounds with more complex conjugated systems have strong and varied absorption between 200–280 nm, involving both π to π^* and n to π^* transitions. Carbonyl compounds such as ketones and aldehydes absorb strongly between 260–300 nm due to n to π^* and π to π^* transitions, with slight shifts in aldehydes caused by the presence of the aldehyde group. Carboxylic acids show a weak to moderate range of absorption (200–230 nm) mainly due to *n* to *π** transitions, which are influenced by conjugation. Amines and phenols have variable and strong absorptions, respectively, spanning 230–300 nm and 210–280 nm, with changes dependent on structure, conjugation, and interactions, such as those between the aromatic ring and hydroxyl group in phenols. Nitro compounds and sulfur compounds also display strong and variable absorption characteristics between 275–400 nm and 230–350 nm, respectively, due to *n* to *π** transitions, with the intensity and breadth influenced by the nitro group in nitro compounds and the electron-donating or electron-withdrawing effects of sulfur in sulfur compounds. This analysis is crucial for understanding the chemical structure and properties of organic molecules by observing how they interact with light at specific wavelengths.

Table 6. Ultraviolet–visible spectroscopy analysis, spectra, and interpretation of organic compounds

Compound Type	λ_{max} (nm)	Electronic Transition	Description of Spectral Features
Alkanes	175–200	*σ* to *σ**	Very weak absorption due to saturated hydrocarbons.
Alkenes	175–195	*π* to *π**	Moderate absorption reflecting *π* to *π** transitions in double bonds.
Aromatic compounds	200–280	*π* to π*, *n* to *π**	Strong absorption; multiple bands due to conjugated system.
Ketones	260–290	*n* to *π**, *π* to *π**	Strong absorption is typical for carbonyl compounds.
Aldehydes	290–300	*n* to *π**, *π* to *π**	Similar to ketones with slight shifts due to the aldehyde group.
Carboxylic acids	200–230	*n* to *π**	Weak to moderate absorption, depending on conjugation.
Amines	230–300	*n* to *π**, *π* to *π**	Variable absorption based on the structure and conjugation.
Phenols	210–280	*π* to *π**, *n* to *π**	Strong, distinct absorption due to aromatic ring and hydroxyl group interaction.
Nitro compounds	275–400	*n* to *π**	Strong absorption, notably broad and intense due to the nitro group.
Sulfur compounds	230–350	*n* to *π**	Variable absorption; is influenced by sulfur's electron-donating or withdrawing effects.

Instrumentation and Technique. A typical UV–Vis spectroscopy instrument consists of a source of light, a monochromator to select the wavelength, a sample cell or cuvette to hold the sample, a photodetector to detect the intensity of transmitted light and a display device that provides the absorbance spectra (Figure 3). The samples are prepared in solution, and the absorbance of light at different wavelengths is measured. This produces a spectrum of peaks at different wavelengths, representing the electronic transitions of the molecule.

Figure 3. Instrumentation and technique for UV–Vis spectroscopy

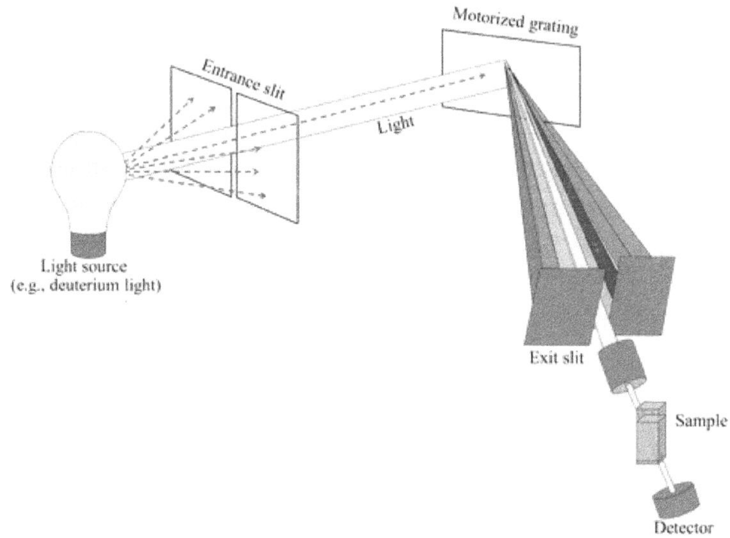

(created by the authors)

Applications in Organic Chemistry. UV–Vis spectroscopy is widely used in organic chemistry for many purposes. The first is the possibility of identifying the sample by comparing its UV–Vis spectrum with that of already known compounds, which is a reliable way of identifying a compound, which is in many cases not possible in all cases with other instruments. The second use of the technique is the determination of concentration; this method perfectly follows the Beer–Lambert law, which is the relationship between the light absorbing ability or absorbance and the concentration of the absorbing species or the solute in solution, making it exact and allowing for simple calculation of concentrations. The third use of the technique is the ability to study the effects of compounds with alternating double bonds, while UV–Vis is particularly efficient for dyes, pigments, and other aromatic compounds. Their spectra reveal the extent of conjugation and how it affects the electronic properties of the compound. Another important use of UV–VIS is its ability to study reactions and mechanisms, as it measures the kinetics and progress of reactions by registering changes in absorbance, and this is an integral part of the study and application of organic chemistry.

Effects of Solvents and Environmental Factors. The solvent and environmental conditions have a significant effect on the UV–Vis spectrum of a compound. Notably, the solvents cause changes in the means of the wavelengths of absorption, while the intensity also varies. These changes tend to occur due to changes in the environment in which the absorbing specimens are situated. Consequently, by using UV–Vis spectroscopy, it becomes possible to determine the solvent effects, pH changes, and environmental conditions that lead to the stimulation of electronic systems of any molecule.

Advanced UV–Vis Techniques. Advances in modern UV–Vis spectrophotometry include the development of the use of UV–Vis systems together with other systems to develop superior approaches for dealing with complex mixtures reliably. Fewer UV–Vis spectrophotometers have also been developed to enable use when small quantities of a specimen are available. The use of time-resolved UV–Vis spectroscopy to monitor reactions in fast biochemical and chemical reactions is also imperative.

Mass Spectrometry

MS is an analytical method that is widely applied in organic chemistry to help acquire molecular species analysis, ascertain structural detail, and visualize complex mixture processes. Moreover, MS offers a versatile and powerful approach for studying chemical reactions, providing real-time monitoring, identification of reaction intermediates and products, elucidation of reaction mechanisms, and quantitative analysis. MS, therefore, ionizes chemical compounds to generate charged molecules or early fragments, which are then assessed in terms of their mass–charge (*m/z*) ratio.

This section offers a concise overview of MS by discussing fundamental principles, techniques, and applications in the field of organic chemistry.

Principles of Mass Spectrometry. By defining the terms ionization mass, mass analysis, and detection, MS is a process that operates on three fundamental principles. The word ionization means converting molecules, meaning that the molecules of a sample are converted to charged ions. This concept is essential because mass spectrometers can only detect charged particles. There are a variety of ways to ionize a sample, such as electron impact and electrospray ionization. The subsequent stage covers the separation of ions based on *m/z* ratios, and this particular principle is known as mass analysis. In experiments, separation is performed with the help of three major types of mass analyzers—quadrupole, time-of-flight, and ion trap analyzers—with different mass resolutions and accuracies. The final phase encompasses the detection of the ions and transformation of the subsequently obtained data into a table of differences in *m/z*. It generally allows for estimations of the molecular weight and proper structures of the backbone with the application of more precise tandem mass spectrometry. These principles and features are why people use mass spectrometry in labs, research, medicine, and environmental studies.

A schematic illustration of a simple mass spectrometer is shown in Figure 4. This shows some essential features of the instrument: an ionization source, a device that separates the produced ions according to their *m/z*, most *m/z* often by magnetic or electric fields, and a detector that counts the number of ions at each *m/z* ratio that generates a mass spectrum. The diagram also shows a vacuum system to keep the ions in a vacuum for charge movement and a data system for spectrum analysis and readout, with every part of the components labeled and the pathway of ions through the gadget indicated. This relatively uncomplicated design illustrates how mass spectrometers detect and calculate the various substances in a sample by comparing the weight of the ions of each molecule.

Figure 4. Schematic representation of a simple mass spectrometer

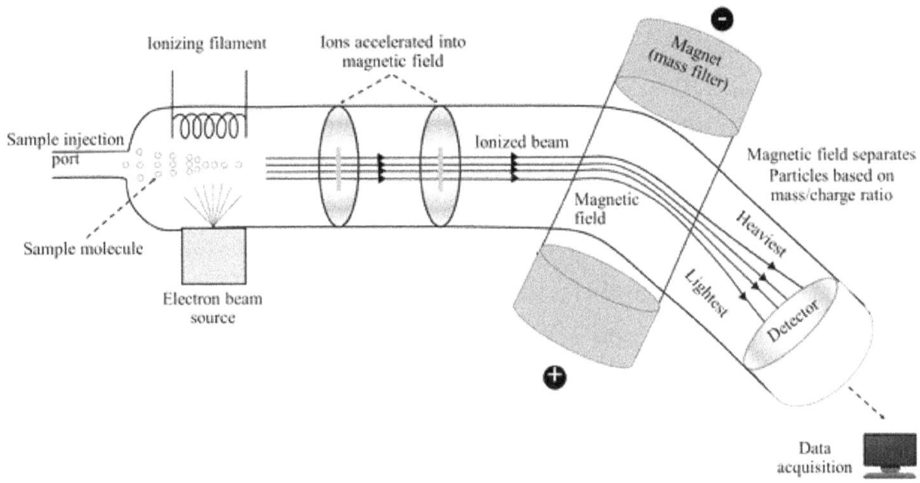

(created by the authors)

Ionization Techniques. Ionization techniques are essential for MS because they are responsible for converting neutral molecules into charged ions that can be analyzed using various detectors. Indeed, different techniques have been developed to enable the analysis of a wide range of samples, from small organic substances to large biomolecules such as proteins. In doing so, numerous methods have been designed and utilized that feature unique mechanisms and application patterns. The quality and efficiency of ionization depend on these mechanisms to meet special demands for each specific analysis. Hence, knowledge of ionization techniques can help optimize experimental conditions, which may improve the accuracy and sensitivity of mass spectrometric analysis. Moreover, various ionization techniques can be used depending on the nature and state of aggregation of the sample. Each ionization method has its own advantages and limitations, and the choice of method depends on the nature of the sample and the specific requirements of the analysis.

Electron impact (EI) is a classical ionization technique that has been widely utilized in MS to analyze volatile and stable organic samples. The ionization approach is based on the bombardment of a vaporized sample by high-energy electrons that are generated by an electron beam. At elevated temperatures, the sample molecules lose one electron, hence transforming into positively charged ions. Moreover, this method is unique in that it generates a specific pattern of carrier fragments that can provide adequate information about the investigated molecule's structure.

Chemical ionization (CI) is a soft ionization technique that complements electron ionization in mass spectrometry, providing an alternative approach for analyzing a wide range of compounds with reduced fragmentation and improved molecular ion detection. The resulting spectra are generally simpler and cleaner, with fewer fragment ions, which can make interpretation more straightforward. It is typically used for volatile and thermally stable compounds. Nonvolatile or thermally labile compounds may not be suitable for this technique.

Electrospray ionization (ESI) has been developed as the most effective technique for analyzing large biomolecules such as proteins and peptides via mass spectrometry. The ionization process occurs by spraying the sample solution with a highly charged needle. Thus, a number of highly diversified fine droplets are formed. These droplets, on the way to the detector, evaporate, leading to the defunct formation of ions. In addition, by generating multiple charged ions, this technique helps reduce the *m/z* ratio and, therefore, enhances the mass spectrometer resolution.

Fast atom bombardment (FAB) is considered a soft ionization technique; therefore, it results in relatively low fragmentation of the sample molecules, preserving the molecular ions. It is suitable for the analysis of nonvolatile and thermally labile compounds and is particularly useful for analyzing polar molecules, such as peptides, proteins, and other biological macromolecules. The sample is dissolved in a suitable solvent, often a glycerol matrix, which helps to absorb the energy from the fast atoms and assists in the ionization process. The fast atoms transfer energy to the sample molecules upon collision, causing ionization without significant thermal degradation.

In addition, matrix-assisted laser desorption/ionization (MALDI) is another mass spectrometry technique most relevant when conducting proteomics and any other research related to large biological molecules. MALDI implies the embedding of the sample in the matrix, which absorbs laser energy. ditch with the laser blueprint over the matrix; therefore, the chemistry help the desorption and ionization of the sample molecules without fragmentation. Hence, the so-called soft ionization method allows for the analysis of the molecule while maintaining its integrity.

Mass Analyzers. Once the charged particles are ionized, they are passed through a mass analyzer to separate the masses of the different ions according to their *m/z*. Mass analyzers differ in their resolution, mass accuracy, and ability to analyze complex mixtures. Some examples of mass analyzers include quadrupole mass analyzers, time-of-flight analyzers, and Orbitraps.

Quadrupole Mass Analyzer. It contains four rod-like electrodes with oscillating electric fields. The electric fields manipulate the path of ions as they pass between the rods. The oscillating field stabilizes ions with specific *m/z* ratios to strike the detector. Due to the selective stabilization, the quadrupole is suitable for filtering ions depending on their *m/z* ratios; hence, it is excellent for fast scanning and separation of ions in a sample.

Time-of-flight (TOF) Analyzer. The analyzer measures the time taken by the ions to move through the vacuum tube over a fixed distance. An ion from ionization is accelerated in the tube by a uniform electric field applied toward the detector. Lighter ions have a smaller *m/z* ratio; hence, they travel quicker than heavy ions; hence, their time of flight is short. The flight time of the ions is then used to measure the *m/z* ratio of each ion; the analyzer is essential for determining the double bond equivalent of a wide range of molecules. In addition, a microTOF is a variant of a time-of-flight mass spectrometer that uses the TOF principle to allow high-resolution mass measurements. In contrast to an evaporative light scattering detector, a microTOF determines the quantity of each ion's time of flight using a precisely vacuum-controlled tube. Moreover, the *m/z* ratios of ions can be determined with extremely high precision and speed. Therefore, it is particularly well suited for analyzing complex mixtures and determining the double-bond equivalents of molecules. The microTOF is employed in innovative research and industrial demand owing to its design and methodological characteristics.

Orbitrap. The analyzer is a loosely trapped high-resolution mass analyzer. In the instrument, the ions are trapped in the electrostatic field within the doughnut-shaped chamber. Ions oscillate around the central axis, and the frequency of oscillation depends inversely on the *m/z* ratio of the ion. The method determines the *m/z* for the ions. It is most suitable for a complex sample and offers higher accuracy values; hence, it is used in scans with high resolution.

According to the authors' experience and literature review, mass analyzers such as quadrupoles, TOF, and Orbitraps each have distinct operating principles and applications in the field of analytical chemistry. Table 7 provides an overview of the operational principles, strengths, and typical applications of three different types of mass analyzers. The quadrupole mass analyzer operates on a principle involving four parallel metal rods, each pair electrically connected and subjected to alternating and direct current voltages, creating a dynamic electric field that selectively permits ions of specific *m/z* ratios to pass, effectively filtering ions during analysis. This analyzer is praised for its high scanning speed, good resolution for small to moderate molecular weights, cost-effectiveness, and robust performance, making it a staple in routine LC–MS applications and drug metabolite studies. In contrast, the TOF analyzer accelerates ions using an electric field to a uniform kinetic energy, sorting them by the time it takes to travel a fixed distance, with lighter ions arriving faster. Its capability for very high mass accuracy and resolution allows it to handle large molecules, such as proteins and polymers, making it ideal for protein identification, peptide mass fingerprinting, and complex proteomics analyses. The Orbitrap analyzer, which captures ions in orbital motion around a central electrode where they oscillate and produce an image current to determine their *m/z* ratios based on oscillation frequency, stands out for its high resolution, mass accuracy, and stability. It supports a wide range of *m/z* values, making it suitable for high-end research applications, metabolomics, lipidomics, and other fields requiring precise mass analysis.

Table 7. Comparison of mass analyzers

Type of Analyzer	Principle	Strengths	Applications
Quadrupole mass analyzer	A quadrupole mass analyzer consists of four parallel metal rods. Each opposing rod pair is connected electrically, and a combination of alternating and direct current voltages is applied. This creates a dynamic electric field that allows only ions of a specific mass-to-charge ratio (*m/z*) to pass through at any given time, filtering out others.	High scanning speed, good resolution for small to moderate molecular weights, relatively low cost, and robust performance.	Widely used in routine LC–MS (liquid chromatography–mass spectrometry) applications, environmental analysis, and drug metabolite studies.
Time-of-flight (TQF) analyzer	In a TOF analyzer, ions are accelerated by an electric field to a kinetic energy that is the same for all ions. The time it takes for the ions to travel a fixed distance then becomes dependent on their *m/z* ratio, with lighter ions arriving faster than heavier ones.	Capable of very high mass accuracy and resolution, can analyze very large molecules like proteins and polymers, and provides true mass measurement.	Used in protein identification, peptide mass fingerprinting, and complex mixture analysis like in proteomics.
Orbitrap	The orbitrap analyzer captures ions in an orbital motion around a central electrode. The ions oscillate along the axis of the electrode, and their image current is measured to determine the *m/z* ratio based on the frequency of oscillation.	High resolution and accuracy, excellent mass stability, and capable of handling a wide range of *m/z* values.	High-end research applications, metabolomics, lipidomics, and other areas where high-precision mass analysis is needed.

The separated ions in the mass analyzer, by interacting with the detector, generate an electrical signal that can be measured, and the data can be processed. A common type of detector is an electron multiplier detector, and in this electron multiplier, the incoming ions strike a dynode; then, the dynode releases an electron that strikes another dynode and releases an electron. This cascade effect then amplifies the signal. Another commonly used detector in mass spectrometry is the Faraday cup detector, which simply collects charged particles and measures current. A mass spectrum is then the file that is produced as output by an MS analysis, and it is essential. Plot of the ion signal intensity with respect to the *m/z* ratio. Each peak in the mass spectrum corresponds to ions with specific *m/z* values.

Then, the height or area of each peak could be directly proportional to the intensity of the ions, and the position in the file could then correspond to the *m/z*. This could help in the identification of the molecular weight of the molecules and the structure of unknown samples. This is at the end of the process, but the main points remain. The last step in the process is interpreting the mass spectrum to deduce the nature of the chemical structure or composition of the sample.

Tandem Mass Spectrometry. Another example is tandem mass spectrometry (MS/MS). It is a modern and highly efficient analytical method that allows for achieving much more detailed and structural insights into molecular composition than can be achieved with single-stage mass spectra. MS/MS includes at least two distantly performed mass analysis steps or stages. Initially, a selected precursor ion from a complex mixture is isolated and fragmented. Subsequently, the resulting product ions are analyzed in subsequent stages. MS/MS technology is instrumental for examining molecular structures in detail and identifying separate substituents, functional groups, and established sequences in biomolecules such as peptides. Thus, the main function of two-stage mass spectrometry is to provide the most accurate information about the structural part of the molecule at the location of the individual fragments. In other words, after breaking into several pieces, the connections between each other and the nature of the connections are established. This detailed information is extremely valuable for the development of targeted drugs, the study of biomolecular mechanisms, or the characterization of the composition and chemical nature of impurities and pollutants.

Table 8 provides an overview of MS analysis and interpretations for several types of organic compounds, highlighting the differences in ionization modes, the appearance of molecular ions, and common fragment ions. For alkanes, positive ionization predominantly results in the loss of methyl (CH_3) and ethyl (C_2H_5) groups, whereas negative ionization typically yields few ions. Aromatic compounds in positive mode commonly form a benzyl ion ($C_7H_7^+$), particularly in mono-substituted benzenes, while negative mode generally results in the loss of a hydrogen atom. Alcohols lose water (H_2O) in positive mode and often lose OH as O^- in negative mode. Carboxylic acids in positive ionization mode lose COOH as CO_2 and H^+, whereas negative ionization mode shows similar losses due to deprotonation. Esters display typical ester cleavage fragments in positive mode and loss of neutral ethyl formate in negative mode. Nitriles exhibit loss of cyanide (CN) in positive mode and minimal fragmentation upon deprotonation in negative mode. Amines in positive mode often lose ammonia (NH_3), and negative ionization is less common due to their low acidity. This detailed analysis helps in deducing the structural components and functional groups of organic molecules based on their fragmentation patterns under different ionization conditions.

Table 8. Mass spectrometry analysis and interpretation of organic compounds

Compound Type	Ionization Mode	Molecular Ion	Common Fragment Ions	Description of Spectral Features
Alkanes	Positive	M^+	M^+–15, M^+–29	The positive mode shows loss of methyl (CH_3) and ethyl (C_2H_5) groups.
	Negative	M^-	Rare	Negative mode generally yields few ions for alkanes.
Aromatic compounds	Positive	M^+	M^+–77	Formation of benzyl ion ($C_7H_7^+$) common in mono-substituted benzenes.
	Negative	M^-	M^-–1	Loss of hydrogen, forming the deprotonated molecule.
Alcohols	Positive	M^+	M^+–18	Loss of water (H_2O) in positive mode typical for alcohols.
	Negative	M^-	M^-–17	Deprotonation often leads to the loss of OH as O^-.
Carboxylic acids	Positive	M^+	M^+–45	Loss of COOH as CO_2 and H^+ in positive mode.
	Negative	M^-	M^-–44	The negative mode favors deprotonation, showing CO_2 loss.
Esters	Positive	M^+	M+–18, M^+–46	The positive mode shows losses typical for ester cleavage.
	Negative	M^-	M^-–59	Loss of neutral ethyl formate ($HCOOC_2H_5$) fragment.
Nitriles	Positive	M^+	M^+–26	Positive mode shows loss of cyanide (CN) fragment.

continued on following page

Table 8. Continued

Compound Type	Ionization Mode	Molecular Ion	Common Fragment Ions	Description of Spectral Features
	Negative	M^-	$M^-–1$	Deprotonation occurs, with minimal further fragmentation.
Amines	Positive	M^+	$M^+–17$	Positive mode often involves the loss of ammonia (NH_3).
	Negative	M^-	Rare	Negative ionization is less common due to low acidity.

Applications in Organic Chemistry. One of the universal and irreplaceable applications of MS in organic chemistry is molecular identification, which is performed using this heroic method. This can be attributed to the molecular weight and formula of organic compounds. This is a complete method for determining the structure of a molecule, which is expressed in structural elucidation by analyzing the fragmentation pattern and creating a map of the molecule from its structural skeleton. It is a quantitative method, as it can measure the concentration of a substance in a sample. Another important use of MS is the study of the mechanisms of reactions: MS can show which intermediates form and which are consumed in the course of a chemical reaction, thereby indicating the dynamics of what is happening in the molecular system. High-resolution and tandem mass spectrometers are particularly useful for detailed reaction studies, providing high accuracy and structural information.

Advances in MS Techniques. Numerous recent technological developments have fundamentally changed the capabilities of MS. Specifically, high-resolution mass spectrometry (HRMS) provides a more convincing approach to determining mass, which is critical for identifying unknown compounds and impurities. Among the notable HRMS advancements, the microtime-of-flight (micro-TOF) mass spectrometer stands out. It utilizes a smaller flight tube, enabling precise mass measurements and high resolution in a compact design. This makes it ideal for rapid analysis and integration with other techniques such as liquid or gas chromatography, thus expanding its utility in areas such as metabolomics and environmental science. New hybrid techniques, which merge numerous analyzers at once, such as quadrupole-time-of-flight (Q-TOF), improve both resolution and broaden horizons, from environmental analysis to pharmacokinetics. Taken together, these spectroscopic methods, utilized separately or simultaneously, constitute a set of tools for structural and reaction analysis of organic molecules and drive the scientific and practical use of chemistry.

KINETIC STUDIES IN MECHANISM ELUCIDATION

Kinetics in chemical sciences refers to measurements of how fast chemical reactions occur and the variables influencing these rates. It strives to measure how the speed is affected by different conditions, including temperature, pressure, concentration, and the presence or absence of catalysts. Kinetics also provides an understanding of the reaction order between the concentrations of reactants and the rate of the reaction idealized by the rate law. The rate law, which is simple or complex based on the number of steps in the reaction mechanism, involves the "what" and "how" chemical reactions occur. The kinetic analysis consisted of the following steps.

Rate Measurement

In chemistry, rate measurement refers to monitoring the degree of change in the concentration of either reactants or products of a given reaction over a period of time. Typically, for this case, methods such as spectroscopy are often used to observe these changes with exactness at regular time intervals. Measuring data from this procedure helps in calculating the rate of the reaction, a typical method for synthesizing aspirin. Figure 5a–b shows the reaction of salicylic acid with acetic anhydride, the principal reaction in the synthesis of aspirin. The reaction between salicylic acid and acetic anhydride always occurs. In particular, the carboxylic acid group of salicylic acid reacts with the anhydride group of acetic anhydride, causing the formation of aspirin and an acid called acetic acid. Notably, in this study, UV–Vis spectroscopy was used because salicylic acid absorbs ultraviolet light at a particular set of wavelengths. As seen, when these light wavelengths are varied, the concentrations of the reactants vary, having an inverse effect, hence offering ideal conditions for practical rate-measuring laboratories. The steps outlined above help in studying what factors increase the speed of the reaction in these experiments. (1) Following this experimental protocol, a calibration curve derived from measured absorbance values is plotted. (2) Under the same controlled conditions and with a regular set of washings, the reaction is currently set, providing samples that can be used to measure absorbances. (3) The calibration curve will then be used to measure the concentration at different times. This data will be used to measure the rate of the reaction. The rate will then be used to analyze the order of the reaction and reaction rate values and how each of these changes with the presence or absence of catalysts or with temperature changes. (4) One must also, using the Arrhenius equation and the discussed methods of data plotting, be able to further determine how varying temperatures led to the abovementioned results. All these conditions lead to a nondestructive, accurate, and relatively simple way of measuring the rate of chemical reactions, such as the rate at which aspirin is produced (Bingham et al., 2024).

Figure 5. Synthesis of aspirin: (a) Typical laboratory chemical synthesis and (b) reaction mechanism

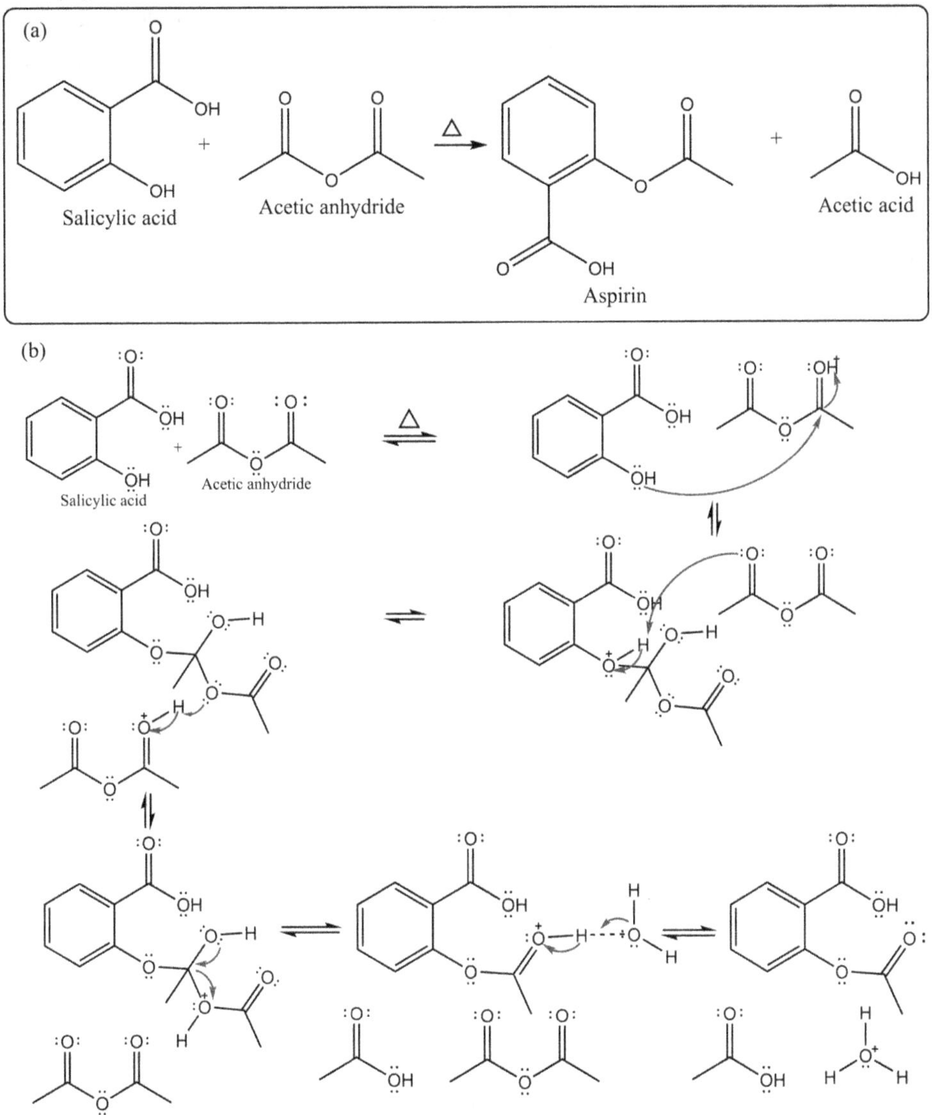

(created by the authors using ChemDraw 20.1.1)

Rate Laws

Rate laws are mathematical equations that indicate how fast a reaction is in terms of the concentration of reactants. Generally, rate laws are derived from experimental data, and they can range from the original direct proportion to complicated variables that involve different types of reactants and conditions. For instance, one of the chemical reactions in the synthesis of a local anesthetic drug, lidocaine, is the reaction between 2,6-dimethylaniline and 2-chloroacetyl chloride (Figure 6). This reaction is an important intermediate reaction because it produces the acylated intermediate required to complete the structure of lidocaine. In this step, 2,6-dimethylaniline, which is an aromatic amine, reacts with 2-chloroacetyl chloride, an acyl chloride. Ideally, a reaction such as the one above requires basic conditions. Basic conditions help the amine deprotonate, unmasking a stronger, more reactive nucleophile, thereby promoting the third step of the reaction mechanism. Then, the nucleophile, the amine group, attacks the carbonyl carbon of 2-chloroacetyl chloride. The carbon-halogen bond is electrophilic in this step, and nucleophilic attack leads to the opening of the acyl chloride, forming an amide bond and eliminating HCl at the same time. This step is amide formation, and amides are intermediates that are further transformed in the fourth step (Silva et al., 2023).

The rate law for this reaction is as follows (Eq. 2):

$$Rate \; = \; k[2{,}6 - dimethylaniline][2 - chloroacetyl\ chloride] \qquad (2)$$

This second-order rate law is applicable in one of the possible scenarios in which the reaction rate is directly proportional to the concentration of each of the two reactants. According to this rate law, the reaction is considered to be the rate-determining step of bimolecular collisions between two reactants. It is pertinent to understand this rate law because it is necessary to optimize the synthesis of lidocaine. The explanation of such a rate law would require the optimal concentration of both 2,6-dimethylaniline and 2-chloroacetyl chloride. This finding also suggested the effect of any dilution on the reaction and other reaction conditions, such as temperature and solvent, which could influence the rate constant k and hence increase the yield.

Figure 6. Synthesis of lidocaine

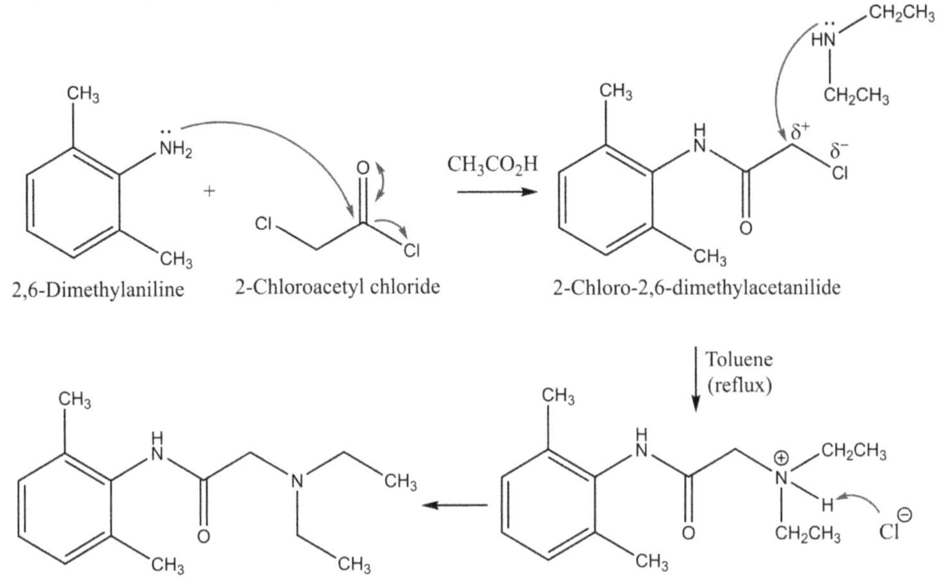

HCl Lidocaine 2-Chloro-2,6-dimethylacetanilide

(Elaborated by the authors using ChemDraw 20.1.1)

Half-life

The half-life is the time it takes for half of the reactant in a chemical reaction to form products. It often indicates how fast or slow a certain drug takes to form reactions and depends on the conditions under which it occurs. One of the many reactions that has implications for the efficiency and stability of penicillin-based antibiotics is the reaction in which penicillin undergoes hydrolysis. Penicilloic acid is essentially an inactive form of penicillin. Nevertheless, the half-life under conditions in which penicillin undergoes hydrolysis may be important in the pharmaceutical industry. The half-life of the reaction is one of the most important since it shows how stable the drug is under varying conditions. For the majority of situations, the general process of penicillin hydrolysis involves the drug breaking down its β-lactam ring to penicilloic acid. pH-temperature enzymes, such as β-lactamase, may have an environmental influence on this process. Briefly, the hydrolysis of penicillin by external influences is the most important time. The most crucial factors are the environment's pH and how it influences the hydrolysis of penicillin. When looking at this specific value, a general rule may apply. The hydrolysis of penicillin is most stable at low pH and less so at high pH. This means that the half-life of penicillin decreases as the pH of the solution increases. The latter means that the half-life of penicillin under acidic conditions is the longest, which means a slower speed of hydrolysis because the opening rate of the β-lactam ring is slower. In contrast, under neutral and alkaline conditions, the pH reaches or exceeds neutral. In the latter case, the half-life of the drug is significantly lower. The likely reason is that the β-lactam ring is much more susceptible to nucleophilic charges in this case, and their rate influences the hydrolysis of penicillin. The half-life of the reaction to the pH can be immense, and the latter needs to be considered when producing the drug. A buffer system would have to be found in the case of pharmaceuticals to maintain the solution's pH value as close as possible to a number that ensures maximum stability. Enteric or other drug coatings could be used so that penicillin is not exposed to an unfavorable environment until it reaches the human equivalent. In some cases, combinations of penicillin with a β-lactamase inhibitor that does not allow the enzyme to hydrolyze penicillin can be made (Lima et al., 2020).

Arrhenius Equation

The Arrhenius equation (Eq. 3) is a chemical formula that reflects the temperature-rate relationship. Moreover, this equation is widely used in chemistry because it defines the rate as being temperature dependent.

$$k = A \times e^{-\frac{E_a}{RT}} \quad (3)$$

where:

k is the rate constant,

A is the preexponential factor or frequency factor, which is a constant for each chemical reaction,

Ea is the activation energy of the action (expressed in joules per mole),

R is the universal gas constant (8.314 J/mol×K),

T is the absolute temperature (in Kelvin),

e is the base of the natural logarithm.

This equation states that an increase in temperature will result in an increase in arrhythmias; thus, the reaction speed increases. The activation energy is one of the important concepts and corresponds to the minimum energy the reagents may have to create the product. The lower the activation energy is, the faster the reaction will occur at a given temperature. This equation is essential in kinetic studies because it allows the rate of reaction to be related to the temperature; thus, the energy required to activate the reaction can be evaluated in a certain way. The identification of the activation palpitate is important because it defines the minimum amount of energy the arrangements may have to convert into palpitations. For instance, an equation may be used to determine the factors affecting the reaction temperature. The following example is given for the creation process of a pharmaceutical component, in which the temperature is modified to optimize the yield. When using the Arrhenius equation, researchers can determine the impact of temperature on the reaction. As a result, it will be possible to determine the best possible conditions in which the reaction rate will be the highest and side reactions will not occur (Kim et al., 2024). Activity palpitation is also important because it defines the energy at which something is needed to achieve the reaction.

ISOTOPE EFFECTS IN SPECTROSCOPY

Isotope Effects

Isotope effects in spectroscopy provide a great medium for understanding the dynamics of chemical reactions, studying reactions when an atom is substituted with one of its isotopes, and determining how it changes the rate or outcome of the reaction. Isotopes are elements of different masses owing to their different neutron counts but maintain the same chemical properties. Common isotopes studied in organic chemistry include deuterium (D) and carbon-13 (^{13}C) isotopes (Glänzer et al., 2024; Roncone et al., 2024). The foundation behind isotope effects is the variations in mass between isotopes, leading to differences in the vibrational frequencies of molecular bonds. These variations have different impacts depending on numerous factors in the reaction, as demonstrated in the following subsections.

Kinetic Isotope Effect (KIE)

The KIE is one of the most significant isotope effects. This difference emerges due to the different masses of isotopes resulting from different vibrational frequencies of the bonds in which they participate. This difference is especially severe in chemical reactions involving isotope bonding or breaking. For instance, replacing a hydrogen atom in a molecule with deuterium can greatly reduce the rate of reaction if the hydrogen is involved in the rate-determining process. This is because it forms stronger bonds due to its greater mass. Chemists can discern which bond is part of the rate-determining step and, hence, deduce more about the reaction mechanism by studying the KIE (Xia & Gao, 2024).

Equilibrium Isotope Effect (EIE)

An EIE is observed when the substitution of an isotope alters the equilibrium constant of the reaction. This is primarily due to the difference between the bonds that isotopes form. Bonds in which heavier isotopes, including deuterium and ^{13}C, form are stronger and more stable. Thus, substituting isotopes can be used in this way to observe the thermodynamic aspects of reactions. Chemists can understand how bond strengths change with isotopic substitution and how these changes affect reaction dynamics (Giraudeau & Remaud, 2020).

Applications of Isotope Labeling in Mechanistic Studies

Isotope labeling in chemical and biochemical research is an approach for studying the mechanism of reactions by tracking atoms through reaction pathways, identifying transient species, and determining the mechanism of action of enzymes. There are several ways in which isotope labeling can be applied in mechanistic studies. They include the following subsections.

Tracing Reaction Pathways

Isotope labeling is a technique that enables chemists to track how and at which position atoms move during reactions between two or more molecules to form a compound. By replacing particular atoms with isotopes in the reactants, they are then able to see where those atoms moved and how each one contributed to the final product. It is particularly helpful and necessary when dealing with complex reactions that can lead to several possible reaction pathways. In synthetic organic chemistry, isotopic tracing explains which pathway is most prevalent and which pathway is activated depending on which conditions or reagents are used (Mouhsine et al., 2024). In this section, the authors explain research on the synthesis of acyclovir, which is an antiviral drug used primarily to treat herpes simplex virus. Acyclovir is synthesized through many chemical steps starting from simpler precursors. Typical synthesis begins with the guanine molecule. One of the key steps in the synthesis pathway is the acylation of guanine by acetic anhydride. Here, guanine is transformed into acycloguanosine. In the synthesis pathway, chemists can label the methyl group of the acyclovir precursor, acyclic guanine, with ^{13}C. They can then observe how the methyl group moves through the reaction stages. By analyzing the behavior of the isotope after its addition at each stage, the most efficient pathways can be determined, and unproductive side reactions or byproducts can be detected. This process includes further chlorination and several reactions that convert it first into a compound and then into acyclovir via several pathways, including the protection and deprotection of the hydroxyl group and direct hydrolysis. The instabilities of guanine bases create the need for protection at certain stages to prevent side reactions. Finally, guanine reacts with the labeled acetic anhydride, which leaves carbon on the oxygen (Guo et al., 2013).

Identifying Intermediates and Transition States

Many chemical reactions involve short-lived intermediates and transition states, which are almost impossible to observe directly. However, in some reactions, deuterium can be intentionally used to 'slow down' the reactions—an approach called isotopic kinetics. In this way, the reaction kinetics are altered to adjust to the level at which intermediates, and transition states accumulate and are then readily detectable. After the accumulation of detectable levels of these species, chemists can use conventional spectroscopic methods, such as NMR or IR spectroscopy, to detect them. Such studies on each intermediate and the structurally related changes that occur one after another in a reaction from the reactants to the products are essential. There, this allows for working out the passing structures and searching to understand their story directly, from reactant to product (Nyagilo et al., 2024). An example of a synthetic pathway where such methods were used is the synthesis of the antipsychotic haloperidol. Isotopic labeling is crucial for stabilizing the intermediate. During the synthesis, the fluorophenyl group must be used, and isotopic labeling is performed to substitute the hydrogens in the shortest periods, where the intermediate can be stabilized. In this way, most commonly only during the introduction of the fluorophenyl group, the hydrogens have to be substituted by deuterium to slow down the reaction, up to the level where the amount of the particular intermediate can be quantified spectrally or quantitatively. The obtained matrices are then subjected to NMR, allowing their structures to be investigated. This is to ensure that the structure of the intermediates is assumed to be proper and efficient. These factors ensure the accuracy of the synthetic pathway and increase the control of haloperidol synthesis (Shamsipur et al., 2007).

Studying Enzyme Mechanisms

A frequent practice in biochemistry is to use isotopes for the study of the mechanisms by which enzymes can facilitate reactions. Enzymes are mostly complex proteins that convert substrates into products. These processes take a specific course and are performed efficiently. However, to observe the precise process occurring in the active site, isotopes may label substrates or enzymes to determine how they interact. Through this approach, researchers can determine whether bonds are broken or formed, and in the case of enzymes, one can identify the active enzyme parts (Yiğit et al., 2023). For example, isotopic labeling is applied when creating statin drugs, of which lovastatin is an example used to aid in lowering high cholesterol levels. Isotopic labeling, particularly using ^{13}C, helps in the study of HMG-CoA reductase in cholesterol biosynthesis. One then labels the carbon atoms in the substrate receiver by the enzyme, which in this case is HMG-CoA, enabling the scientists to trace the enzymatic reaction process converting to mevalonate, an essential element in cholesterol synthesis. The exact response occurring in the enzyme active site can then be observed, as this insight helps in designing better drugs that can effectively stop the enzyme (Su et al., 2024).

Quantitative Analysis

Isotopic labeling of compounds is an essential technique in quantitative analysis for analytical techniques such as MS and NMR spectroscopy. Stable isotopes are used as internal standards in these measurements and serve as a reference for the amount of an unknown compound in a mixture. The application of isotopes guarantees the precision and reproducibility of the measurements and, given the complexity of analytical mixtures, is an inherent requirement in analytical chemistry (Mouhsine et al., 2024). A good example in drug synthesis where isotopic labeling serves quantitative purposes is ^{13}C-methotrexate in pharmaceutical studies. Methotrexate is a chemotherapeutic immunosuppressive agent and a compound whose quantification in biological matrices may necessitate pharmacokinetic studies. It can be synthesized through an aminopterin reaction with formaldehyde and a reducing agent—normally sodium borohydride. In this synthesis, the effect of reduction is the reduction of the amide group to an amine. This process involves the conversion of a relative of aminopterin, also called methopterin, to methotrexate (Kovalev et al., 2022). When using isotopically labeled formaldehyde, e.g., ^{13}C-formaldehyde, the course of this conversion may be limited. This method of comparing the rate of incorporation of a methyl group into the structure or not of methotrexate allows tracking and accurate "measurement" of the drug in the biological system, which is a designation of this marker for quantitative goals in pharmacokinetic studies (Song et al., 2024).

Table 9 provides insight into how isotopic labeling can be introduced into different chemical reaction mechanisms. Ogo et al. (2024) reported a kinetic isotope effect of ^{13}C in the hydrogenation of alkenes, indicating that the rate-determining step of the mechanism appears to involve

C–H bond formation. The oxidation of alcohols using ^{18}O in the work of Tateishi et al. (2024) is an example in which the use of ^{18}O revealed a change in the product distribution, suggesting that oxygen is involved in the rate-limiting step. In addition, if the solvent is changed and when the solvolysis reactions are performed using ^{2}H (deuterium), a clear decrease in the reaction rate is observed, Dorsheimer and Rovis (2023) which is consistent with C–D bond formation in the transition state. These cases undoubtedly illustrate the usefulness of isotopic labeling as a powerful tool in enabling a more extensive understanding of what changes are being affected at the molecular level, resulting in the observed trends of reaction pathways and rates and therefore valuable insight into the mechanistic details of these chemical processes.

Table 9. Isotopic labeling results and mechanistic insights

Isotope Used	Reaction Investigated	Key Findings	Mechanistic Insights	References
^{13}C	Hydrogenation of alkenes	Shift in rate with ^{13}C	The isotope effect suggests a rate-limiting step involving C–H bond formation	Ogo et al. (2024)
^{18}O	Oxidation of alcohols	Altered product distribution with ^{18}O	Indicates the involvement of oxygen in the rate-limiting step.	Tateishi et al. (2024)
^{2}H (deuterium)	Solvolysis reactions	Slower reaction rate with deuterium	Supports the mechanism of C–D bond breaking in the transition state.	Dorsheimer and Rovis (2023)

COMPUTATIONAL METHODS IN MECHANISM ELUCIDATION

Computational Techniques

Various computational techniques in chemistry are essential resources for the elucidation of mechanisms because they can mimic chemical reactions between molecules. Computational methods utilize theoretical models and extensive calculations, enabling the prediction of molecular behavior under different chemical stimuli. Additionally, such methods can reveal information on the dynamics of reactions, which is often difficult to obtain using experimental methods (Ismail et al., 2024). Some of the key computational methods for elucidating the underlying mechanisms are as follows: quantum mechanics – this method allows the modeling of molecules' electronic structures using basic quantum mechanics principles. Quantum mechanics (QM) uses methods such as density functional theory and Hartree-Fock calculations to study reaction mechanisms and transition states between molecular structures. Molecular Mechanics (MM) – The MM method is based on classical physics and uses interactions between atoms and molecular entities to model systems. MM is typically used for larger molecules, where quantum mechanics methods become impractical due to excessive computational time requirements. Molecular dynamics – The simulation of molecules over time reveals their movement, dynamics, and interaction. Molecular dynamics uses simulations to understand molecular dynamics and the impact of conformation on chemical reactions, including reactions in solutions or biological systems. Hybrid QM/MM methods – the combination of strength from QM precision with QM/MM fast-making computations is often utilized to study enzymes or heterogeneous systems. Nevertheless, all of the methods presented can predict the activity, selectivity, and rate of the process to a certain extent, and all of them are vital tools for modern chemists to study various chemical processes (Sastre et al., 2024).

Integration of Computational Methods with Spectroscopic Data

Integrating computational methods with spectroscopic data is a highly effective and synergistic approach to increasing the level of detail in the understanding of chemical reactions. Understanding and improving chemical reactions is a two-step process involving two types of drier knowledge: theoretical and one originating from rapid experimentation. It is commonly assumed that there is a discrepancy between the two types of knowledge. However, they are two sides of the same shiny coin. The proper attitude toward this is outlined in the last integration point: the most important compounds to include in a chemical substance. Theoretically, these intermediates are involved. Noiselessly, these are reagents and final compounds that, when spectated, residue as possible intermediates. These and others can be affected by this because, in this way, their compounds residue and soundly about the limit of chemical substances. Both work in tandem, allowing both to gain and sustain points (Ismail et al., 2024).

ADVANCES AND EMERGING TRENDS

In this section, the authors discuss the leading research directions in spectroscopy to illustrate how recent technological progress, evolving capabilities, and expected future developments determine the field's progress. According to the authors, examining recent developments not only allows for the presentation of current refined experiments but also enables predictions of future impacts on scientific inquiry and experimentation.

Recent Technological Advances in Spectroscopy

The field of spectroscopy has been greatly influenced by advances in techniques that extend their capabilities due to new technologies that can be applied to study organic reaction mechanisms. Recent advances in the sensitivity, resolution, and speed of various spectroscopic techniques help characterize the chemical nature of analytes more accurately and quickly. These advances include the following.

First, the development of ultrafast spectroscopy allows for the observation of a single chemical reaction, studies on the timescale of femtoseconds (10^{-15} seconds), and tracking of the dynamics of electron movements and bond formation. This technology sheds new light on what was previously a transient state and an unclear reaction pathway. Owing to the use of ultrashort laser pulses, these images demonstrate both the rapid motion of electrons and the formation of chemical bonds. This is important because slower methods are unable to capture such immediate and transient states. Ultrafast spectroscopy provides a means to recognize and evaluate these compounds. It is crucial due to its potential applications in areas such as photochemistry, molecular dynamics analysis, and work on new materials, where rapid processes at the atomic and molecular levels regularly occur thereafter (Jang et al., 2024).

Second, in the area of organic chemical synthesis, 2D-NMR spectroscopy plays a critical role as an analytical technique that provides a key that allows chemists to ascertain the configuration of complex molecular structures. Understanding the molecular structure of complicated organic molecules and biomolecules is crucial. The 2D-NMR techniques COSY, NOESY, and HSQC give chemists the information needed to deduce which two atoms are associated, resulting in obtaining the three-dimensional structure of a molecule. In particular, the potential of the 2D-NMR technique makes it feasible to determine how the novel molecules synthesized are arranged. Additionally, it allows for the comprehension of intricate biochemical phenomena based on the association of confirmation and/or organic molecule structure (Morris, 2024).

Third, new types and approaches in Raman spectroscopy – SERS (surface-enhanced Raman spectroscopy) and TERS (tip-enhanced Raman spectroscopy) – have allowed scientists to detect minuscule species and use Raman spectroscopy for surface chemistry and catalysis. When regular Raman spectroscopy measurements require at least nanomolar concentrations of the molecule to have a detectable Raman signal, the enhanced versions allow for sensitive detection and analysis of the matter at the single-molecule level. The SERS technique operates by putting the target molecules on or near a roughened metal surface or metallic nanoparticles such as silver, gold, or copper, which significantly enhances the intensity of the Raman signal due to the local electric field of the light-activated metal and is capable of detecting a single molecule. TERS, on the other hand, incorporates the above principles with high-resolution surface imaging technology for scanning probe microscopy. TERS utilizes a metallic sharp-edged tip in close proximity to the sample surface. The tip, acting like an antenna when exposed to light, enhances the local intensity of the electric field on its top. As a result, Ramin-active molecules localized in this area emit a significantly more intense Raman signal, which also dramatically increases the resolution of Raman spectroscopy. Consequently, the SERS and TERS techniques are widely adopted in chemistry, materials science, and biology due to their ability to acquire molecular information and images without extensive sample preparation (Murugasenapathi & Palanisamy, 2024).

Future Directions: Machine Learning and Real-Time Monitoring

Based on recent literature and the authors' knowledge, the use of machine learning in spectroscopic data analysis is one of the current hot topics due to its considerable potential for evolution. Machine learning algorithms can process large data volumes obtained from spectroscopy-driven experiments and, based on that data, derive patterns and correlations that are difficult to capture via traditional data-processing methods. Using machine learning, spectroscopists can make more accurate predictions of specific chemical properties and behaviors, identify new compounds, and optimize reaction conditions. Machine learning is beneficial for spectroscopic data processing, and it can also be utilized for in silico spectroscopy development strategies by enabling better-structured illumination and calibration of instruments (Yu et al., 2024). In addition, data driving can be especially important for the development of real-time monitoring techniques. Currently, several techniques for online NMR, real-time IR, and continuous UV–Vis spectroscopy have been developed. These tools allow for observing the reaction in progress and receiving immediate feedback on the reaction progress, mechanism, and possible reaction kinetics. Thus, real-time spectroscopy is beneficial in industrial settings for generating highly concentrated reaction flows. Additionally, in synch with chemistry, such spectra can be further analyzed within a certain time frame to optimize the reaction according to the obtained data (Caceres-Cortes et al., 2024).

CONCLUSION AND FUTURE SCOPE

This chapter discusses the crucial role of spectroscopic methods in detecting organic reactions. The major findings indicate the diversity and complexity of each spectroscopy technique, including NMR, IR, UV–Vis, and MS, and how each technique reveals different aspects of molecular structure, dynamics, and responsiveness. This chapter elaborates on kinetic studies and isotope effects and discusses how the two study approaches help to better understand the reaction mechanism. Additionally, the chapter discusses the combination of computational and spectroscopy topics and emphasizes how these two topics combine to improve the perspective and accuracy of mechanistic studies. The authors mention technology enhancement machine learning and real-time monitoring as two major contributors that are reshaping the study process of chemical pathways. In conclusion, this chapter emphasizes that spectroscopy techniques have aided organic chemistry by providing necessary tools for exploring and understanding organic chemical reactions at the molecular level. Indeed, it has made a significant contribution to the discovery and identification of reaction mechanisms and the development of synthetic strategies and materials. This chapter highlights that they have attained a perfect level of observing how molecules interact and change, thereby enabling advances in chemical development by increasing efficiency and sustainability. Furthermore, spectroscopy has bridged experimental and theoretical domains, enabled the validation and refinement of computational models, and promoted a more comprehensive understanding of chemical phenomena.

The future of spectroscopy in organic chemistry will entail increased consolidation of computational methods, spectral strategies, and digital technology. The authors anticipate that these methods will take a variety of types, ranging from enhanced algorithms to intelligent real-time methods that can rapidly evaluate challenging data and draw up new findings to become more prevalent. Additionally, the authors anticipate growth in the development of each new spectral method, likely with greater resolution and greater sensitivity, which is necessary for the latest advancements. The enhancements will enable the observation of more details about chemical reactions, further organic evolution, and the discovery of new opportunities. Hence, spectroscopy is an essential part of this field.

ACKNOWLEDGMENT

The authors are grateful to the reviewers and editor for their valuable suggestions, which significantly improved the quality of this chapter.

REFERENCES

Anderson-Wile, A. M., Hall, K. E., & Celius, T. C. (2024). Scaffolded structure determination activities utilizing 2D NMR Spectroscopy of simple organic molecules. *Journal of Chemical Education, 101*(3), 1356–1360. DOI: 10.1021/acs.jchemed.3c00882

Bingham, N. M., Wright, J. S., Mathias, S. C., Douce, D., & Sears, P. (2024). Atmospheric solids analysis probe mass spectrometry: An easy bolt-on for the synthetic undergraduate teaching laboratory. *Rapid Communications in Mass Spectrometry, 38*(5), e9689. DOI: 10.1002/rcm.9689 PMID: 38248910

Bromfield Lee, D., & Nelson, J. (2024). Modifying an organic chemistry esterification teaching lab to be accessible to blind and visually impaired (BVI) students. *Journal of Chemical Education, 101*(5), 1891–1898. DOI: 10.1021/acs.jchemed.3c01081

Caceres-Cortes, J., Falk, B., Mueller, L., & Dhar, T. G. M. (2024). Perspectives on nuclear magnetic resonance spectroscopy in drug discovery research. *Journal of Medicinal Chemistry, 67*(3), 1701–1733. DOI: 10.1021/acs.jmedchem.3c02389 PMID: 38290426

Dorsheimer, J. R., & Rovis, T. (2023). Late-stage isotopic exchange of primary amines. *Journal of the American Chemical Society, 145*(44), 24367–24374. DOI: 10.1021/jacs.3c09442 PMID: 37889497

Giraudeau, P., & Remaud, G. (2020). Advanced NMR Methods for Position-Specific Isotope Analysis. In Yoshida, N., Gilbert, A., & Foriel, J. (Eds.), *Handbook of Isotopologue Biogeochemistry* (pp. 1–31). Springer Nature Singapore.DOI: 10.1007/978-981-10-7048-8_12-1

Glänzer, D., Pfeiffer, M., Ribar, A., Zeindl, R., Tollinger, M., Nidetzky, B., & Kreutz, C. (2024). Efficient synthetic access to stable isotope labeled pseudouridine phosphoramidites for RNA NMR Spectroscopy. *Chemistry (Weinheim an der Bergstrasse, Germany)*, e202401193. DOI: 10.1002/chem.202401193 PMID: 38652483

Guo, H. M., Wu, S., Niu, H. Y., Song, G., & Qu, G. R. (2013). Chemical Synthesis of Acyclic Nucleosides. In *Chemical Synthesis of Nucleoside Analogs* (pp. 103–162). Wiley., DOI: 10.1002/9781118498088.ch3

Hoenig, S. L. (2024). *Basic Chemical Concepts and Tables* (2nd ed.). CRC Press. DOI: 10.1201/9781003396512

Ismail, F. M. D., Nahar, L., & Sarker, S. D. (2024). Prediction of structure based on spectral data using computational techniques. In Sarker, S. D., & Nahar, L. (Eds.), *Computational Phytochemistry* (2nd ed., pp. 255–300). Elsevier. DOI: 10.1016/B978-0-443-16102-5.00016-X

Jang, J., Park, J., & Shin, H. J. (2024). Terahertz spectral analysis: An in-depth exploration of spectroscopy approaches for ultrafast dynamics. *Current Applied Physics*, *59*, 197–207. DOI: 10.1016/j.cap.2023.11.015

Jirasek, M., Sharma, A., Bame, J. R., Mehr, S. H. M., Bell, N., Marshall, S. M., Mathis, C., MacLeod, A., Cooper, G. J. T., Swart, M., Mollfulleda, R., & Cronin, L. (2024). Investigating and quantifying molecular complexity using assembly theory and spectroscopy. *ACS Central Science*, *10*(5), 1054–1064. DOI: 10.1021/acscentsci.4c00120 PMID: 38799656

Kim, J., Hayashi, Y., Badr, S., Okamoto, K., Hakogi, T., Furukawa, H., Yoshikawa, S., Nakanishi, H., & Sugiyama, H. (2024). Kinetic study and model-based design space determination for a drug substance flow synthesis using an amination reaction via nucleophilic aromatic substitution. *Organic Process Research & Development*, *28*(5), 1793–1805. DOI: 10.1021/acs.oprd.3c00380

Koothradan, F. F., Jayarani, A., & Sivasankar, C. (2024). Synthesis of β-lactams through carbonylation of diazo compounds followed by the [2 + 2] cycloaddition reaction. *The Journal of Organic Chemistry*, *89*(7), 4294–4308. DOI: 10.1021/acs.joc.2c01484 PMID: 36469500

Kovalev, I. S., Zyryanov, G. V., Santra, S., Majee, A., Varaksin, M. V., & Charushin, V. N. (2022). Folic acid antimetabolites (antifolates): A Brief review on synthetic strategies and application opportunities. *Molecules (Basel, Switzerland)*, *27*(19), 6229. DOI: 10.3390/molecules27196229 PMID: 36234766

Lima, L. M., Silva, B. N. M., Barbosa, G., & Barreiro, E. J. (2020). β-lactam antibiotics: An overview from a medicinal chemistry perspective. *European Journal of Medicinal Chemistry*, *208*, 112829. DOI: 10.1016/j.ejmech.2020.112829 PMID: 33002736

Linne, M. A. (2024). *Spectroscopic Measurement: An Introduction to the Fundamentals*. Elsevier., DOI: 10.1016/C2022-0-00036-2

Ma, J., Ye, Q., Green, R. A., Gurak, J., Ayers, S., Huang, Y., & Miller, S. A. (2024). Overcoming NMR line broadening of nitrogen containing compounds: A simple solution. *Magnetic Resonance in Chemistry*, *62*(3), 198–207. DOI: 10.1002/mrc.5432 PMID: 38258438

Morris, A. L. C. (2024). Nuclear magnetic resonance spectroscopy in drug discovery, In the previous edition, W.F. Reynolds contributed to this chapter. In McCreath, S. B., & Clement, Y. N. (Eds.), *Pharmacognosy* (2nd ed., pp. 649–684). Academic Press. DOI: 10.1016/B978-0-443-18657-8.00004-9

Mouhsine, B., Norlöff, M., Ghouilem, J., Sallustrau, A., Taran, F., & Audisio, D. (2024). Platform for multiple isotope labeling via carbon–sulfur bond exchange. *Journal of the American Chemical Society*, *146*(12), 8343–8351. DOI: 10.1021/jacs.3c14106 PMID: 38498972

Murugasenapathi, N. K., & Palanisamy, T. (2024). Electrochemical Surface-Enhanced Raman Spectroscopy (EC-SERS): Techniques, Applications, and Future Perspectives. In Singh, D. K., Kumar Mishra, A., & Materny, A. (Eds.), *Raman Spectroscopy: Advances and Applications* (pp. 199–214). Springer Nature Singapore. DOI: 10.1007/978-981-97-1703-3_9

Nyagilo, V. O., Mallojjala, S. C., & Hirschi, J. S. (2024). Transition state analysis of key steps in dual photoredox-cobalt-catalyzed elimination of alkyl bromides. *ACS Catalysis*, *14*(7), 4683–4689. DOI: 10.1021/acscatal.3c06324 PMID: 39211423

Ogo, S., Yatabe, T., Miyazawa, K., Hashimoto, Y., Takahashi, C., Nakai, H., & Shiota, Y. (2024). Cyclopropanation using electrons derived from hydrogen: Reaction of alkenes and hydrogen without hydrogenation. *JACS Au, 4*(4), 1615–1622. DOI: 10.1021/jacsau.4c00098 PMID: 38665645

Osifová, Z., Socha, O., & Dračínský, M. (2024). NMR-Challenge.com: Exploring the most common mistakes in NMR assignments. *Journal of Chemical Education, 101*(6), 2561–2569. DOI: 10.1021/acs.jchemed.4c00092

Roncone, A., Kelly, S. D., Giannioti, Z., Hauk, C., Caillet, C., Newton, P. N., Perez-Mon, C., & Bontempo, L. (2024). Stable isotope ratio analysis: An emerging tool to trace the origin of falsified medicines. *Trends in Analytical Chemistry, 174*, 117666. DOI: 10.1016/j.trac.2024.117666

Sancéau, J. Y., Maltais, R., Zhou, M., Lin, S. X., & Poirier, D. (2024). Synthesis and characterization of targeted 17β-hydroxysteroid dehydrogenase type 7 inhibitors. *The Journal of Steroid Biochemistry and Molecular Biology, 242*, 106544. DOI: 10.1016/j.jsbmb.2024.106544 PMID: 38754521

Sastre, S., Manta, B., Semelak, J. A., Estrin, D., Trujillo, M., Radi, R., & Zeida, A. (2024). Catalytic mechanism of mycobacterium tuberculosis methionine sulfoxide reductase A. *Biochemistry, 63*(4), 533–544. DOI: 10.1021/acs.biochem.3c00504 PMID: 38286790

Shamsipur, M., Shafiee-Dastgerdi, L., Talebpour, Z., & Haghgoo, S. (2007). 19F NMR as a powerful technique for the assay of anti-psychotic drug haloperidol in human serum and pharmaceutical formulations. *Journal of Pharmaceutical and Biomedical Analysis, 43*(3), 1116–1121. DOI: 10.1016/j.jpba.2006.09.038 PMID: 17084055

Silva, A., Mourão, J., & Vale, N. (2023). A review of the lidocaine in the perioperative period. *Journal of Personalized Medicine, 13*(12), 1699. DOI: 10.3390/jpm13121699 PMID: 38138926

Song, Y., Feng, C., Zhou, D., Ma, Z., He, L., Zhang, C., Yu, G., Zhao, Y., Yang, S., & Xing, X. (2024). Constructing efficient bacterial cell factories to enable one-carbon utilization based on quantitative biology: A review. *Quantitative Biology, 12*(1), 1–14. DOI: 10.1002/qub2.31

Su, X. Z., Zhang, L. F., Hu, K., An, Y., Zhang, Q. P., Tang, J. W., Yan, B.-C., Li, X.-R., Cai, J., Li, X.-N., Sun, H.-D., Jiang, S.-Y., & Puno, P. T. (2024). Discovery of natural potent HMG-CoA reductase degraders for lowering cholesterol. *Angewandte Chemie International Edition, 63*(6), e202313859. DOI: 10.1002/anie.202313859 PMID: 38055195

Tateishi, Y., McCarty, K. D., Martin, M. V., & Guengerich, F. P. (2024). Oxygen-18 labeling defines a ferric peroxide (compound 0) mechanism in the oxidative deformylation of aldehydes by cytochrome P450 2B4. *ACS Catalysis, 14*(4), 2388–2394. DOI: 10.1021/acscatal.4c00106 PMID: 38384943

Venetos, M. C., Elkin, M., Delaney, C., Hartwig, J. F., & Persson, K. A. (2024). Deconvolution and analysis of the ^{1}H NMR spectra of crude reaction mixtures. *Journal of Chemical Information and Modeling, 64*(8), 3008–3020. DOI: 10.1021/acs.jcim.3c01864 PMID: 38573053

Xia, X., & Gao, Y. (2024). Compound specific, intramolecular, and clumped 13C fractionations in the thermal generation and decomposition of ethane and propane: A DFT and kinetic investigation. *Geochimica et Cosmochimica Acta, 375*, 50–63. DOI: 10.1016/j.gca.2024.05.001

Yiğit, B., Taslimi, P., Barut Celepci, D., Taskin-Tok, T., Yiğit, M., Aygün, M., Özdemir, İ., & Gülçin, İ. (2023). Novel PEPPSI-type N-heterocyclic carbene palladium(II) complexes: Synthesis, characterization, in silico studies and enzyme inhibitory properties against some metabolic enzymes. *Inorganica Chimica Acta, 544*, 121239. DOI: 10.1016/j.ica.2022.121239

Yu, L., Chen, B., Li, Z., Su, Y., Jiang, X., Han, Z., Zhou, Y., Yan, D., Zhu, X., & Dong, R. (2024). Computer-aided automated flow chemical synthesis of polymers. *Giant (Oxford, England), 18*, 100252. DOI: 10.1016/j.giant.2024.100252

Chapter 6
Synthesis of Sulfur and Sulfoxides Organic Compounds Through Rearrangement Reaction:
Approaches to Drug Design and Synthesis

Zuber Peermohammed Shaikh
https://orcid.org/0009-0007-2606-8459
University of Pune, India

ABSTRACT

Sulfur imidations can be addressed by several synthetic approaches, depending on nature of the starting sulfide or sulfoxide. A wide range of imidating agents has been developed including N-haloamides and related reagents, azides, heterocyclic nitrene precursors, and iodinanes. All are characterised by their specific properties and reactivities. Most of the imidation procedures have been developed for sulfide imidation, while only a few are efficient for sulfoxide conversions. This is particularly true for metal-free variants of imidation reactions. Importantly, the nature of nitrogen-protecting group. This chapter demonstrates synthesis of sulfon-

DOI: 10.4018/979-8-3693-6473-4.ch006

amide using amines and thiol through different oxidative and non-oxidative rearrangement reactions. The significance of sulfoxides as commodity chemicals, their efficacy as versatile building blocks for the assembly of chemically and biologically useful molecules, including therapeutic agents, is increasing enantiopure-sulfoxides, owing to their important biological activities are highly useful in pharmaceutical industry.

INTRODUCTION TO SYNTHESIS OF SULFUR AND SULFOXIDES THROUGH DRUG DISCOVERY

Green chemistry presents a new challenge for organic synthesis, reducing volatile solvent emissions and harmful chemicals. It lessens the usage of hazardous chemicals and the emission of volatile organic solvents. The current study's goal has also employed environmentally friendly oxime production techniques. The traditional method of synthesizing oximes involves the use of an acid catalyst and occasionally refluxing an amine and aldehyde mixture in an organic medium. Introduction Sulfur has been known for centuries, and it is widely distributed in nature. Owing to the presence of sulfur in all living organisms as well as vegetables in the form of organic or inorganic compounds, sulfur is regarded as an important biogenic element. Sulfoxides are important organosulfur compounds that have been extensively used in organic synthesis. This sulfoxide plays a significant role in medicinal chemistry. Several classes of medicinally important drugs like Proton Pump Inhibitors (Omeprazole, Pantoprazole), Anti-psychotics (Modafanil), anti-inflammatory (Suindec) contain sulfoxide as a functional group. Many sulfur containing drugs get metabolised by converting to sulfoxides in the first step of the metabolism pathway. Sulfoxidation plays a crucial role in regulating the tautomeric balance, which ultimately leads to the rapid racemisation of thiazolidinedione class of drugs used for the treatment of diabetics. In the metabolic pathway, the generation of sulfoxide intermediates can produce toxic effects. These key sulfoxide intermediates are responsible for toxicity and can be best explained by quantum chemical studies. We have explored the generation of the sulfoxide intermediates in metabolism of different chemical class of drugs such as thiophenes, thiazoles, thiazolidinediones ex. through quantum chemical analysis, (Colobert F., 1998).

The first step in sulfur imidation is to choose the adequate imidating agent and suitable reaction conditions. Several questions have to be addressed before choosing the optimal procedure: is the substrate compatible with acid/basic conditions, or are neutral reaction conditions essential? Is it easier to isolate the corresponding imidated sulfur derivative as a salt or as a neutral product? Is a transition metal catalyst needed? How about safety and toxicity issues of the imidating agents? Interestingly, all imidating agents (R-N-LG) depicted, have a recurrent structure consisting of first, a nitrogen atom (N) needed for the S–N bond, second, a remaining fragment (R) linked to the nitrogen before and after imidation which can be considered as a protecting group, and, third, a leaving group (LG) which is removed during the reaction. For a better understanding of the advantages and disadvantages of each imidation method, we have classified the examples reported in the literature according to the nature of the imidating agent: N-haloamides and associated derivatives such as chloramine-T and MSH, heterocyclic nitrene sources including oxaziridines and 1,4,2-dioxazol-5-ones, azides and iminoiodinane derivatives. This choice aims to highlight the general behaviour of such reagents in reactions with sulfides and/or sulfoxides by showing their reactivity, the substrate scope, the supposed mechanisms, and their use in stereoselective sulfur imidations. This tutorial review cannot be exhaustive, but our goal is to provide the reader a practical ''toolbox'' discussing the main achievements in the development of efficient sulfide and sulfoxide imidation methods. As we recently devoted a tutorial review to the syntheses, properties and applications of fluorinated sulfoximines, imidations of fluorinated sulfides and/or sulfoxides will not be addressed in this manuscript (Carreño M. C., 2002).

Figure 1. Structures of few medicinally privileged sulfoxides

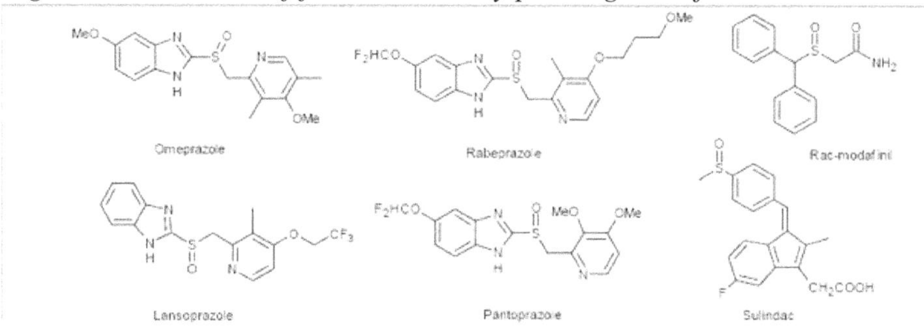

LITERATURE REVIEW OF ORGANIC COMPOUNDS THROUGH REARRANGEMENT REACTION

N-Haloamides and associated reagents (RNH-LG) The common feature of N-haloamides and related compounds is that the nitrogen to be transferred is connected to a leaving group (a halogen or a RSO3 moiety) and that the imidation process involves the formation of a sulfonium or sulfiliminium salt. These reagents are exclusively used for nitrogen transfer onto sulfides providing sulfimides.

Figure 2. Sulfoxide containing drugs in medicinal chemistry

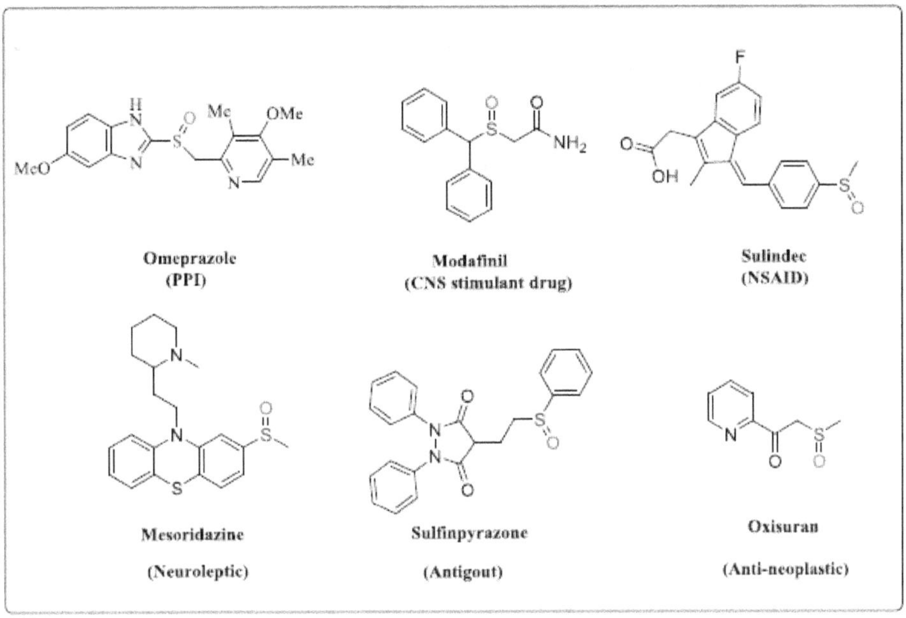

Imidation of Sulfides with N-Chloro Sulfonamides

The first imidation reagent ever used in conversions of sulfides into sulfimides was chloramine-T (N-chloro tosylamide, sodium salt, Using dry conditions with methanol as solvent at a pH of around (adjusted by the addition of acetic acid) gave access to a variety of N-tosyl sulfimidesin moderate to high yields. When water was present, the yields were lower due to the competing formation of the corresponding sulfoxides. The proposed mechanism involves a protonation of 5 to give 13 that serves as chlorinating agent for sulfide. The latter step is slow and rate-determining generating highly reactive sulfonium chloride in combination with tosyl amidate. Both ionic species react quickly with each other yielding by HCl elimination. Replacing methanol by acetonitrile as solvent proved to be highly beneficial. Simple stirring of the sulfides within acetonitrile led to the corresponding N-sulfonyl sulfimides in high to excellent yields (up to 99%). Even in the presence of water, the formation of sulfoxides has never been observed, making this procedure more convenient by using the hydrated form. The absence of sulfoxide led to doubts related to the proposed mechanism as it suggested that the reaction did not proceed through intermediacy. This notion was supported by experiments with enantiopure 2-alkylthianes. Forming the corresponding sulfonium chloride with t-BuOCl and treatment of this intermediate with preformed tosyl amidate (conditions A,) led to the expected syn selectivity (anti/syn = 10: 90), resulting from an anti-electrophilic chlorination followed by an SN2-type imidation. Inverting the addition order of the reagents (conditions A0) afforded the anti product predominantly, presumably by direct NTs imidation. Interestingly, imidation of 16 with 5 in both methanol (conditions B)8 and acetonitrile (conditions C)7 resulted in the same anti selectivity (anti/syn = 90: 10) implying a direct NTs transfer to the less-hindered face of the sulfide under these conditions as well. Thus, both the absence of sulfoxide formation under non-anhydrous reaction conditions and the anti selectivity rendered the initially proposed mechanism (Patai, 1988).

Imidation of Sulfides with N-Haloamides

Reactions between sulfides 1 and ethyl N chlorocarbamate afforded iminosulfonium chloride salts 20a in moderate yields. Treatment of these salts with triethylamine provided the corresponding N-ethoxycarbonylsulfimides in high yields (80–90%). Similarly, N-carbonylsulfimides were obtained by using N-bromoacetamide or N-chlorourea. Imidation with 6b proceeded only with dialkylsulfides affording sulfimides in yields ranging from 36 to 71% over two steps. The reaction failed, when applied to bulky dialkyl or aromatic sulfides. In contrast, with imidations of aryl-alkyl sulfides worked well leading to the corresponding products yields between 42%and 68%after a basic workup. No reaction occurred with diarylsulfides. The mechanism was proposed to be similar to the one described for imidations with chloramine-T.

Figure 3. Imidations of sulfides with N-chloro sulfonamides

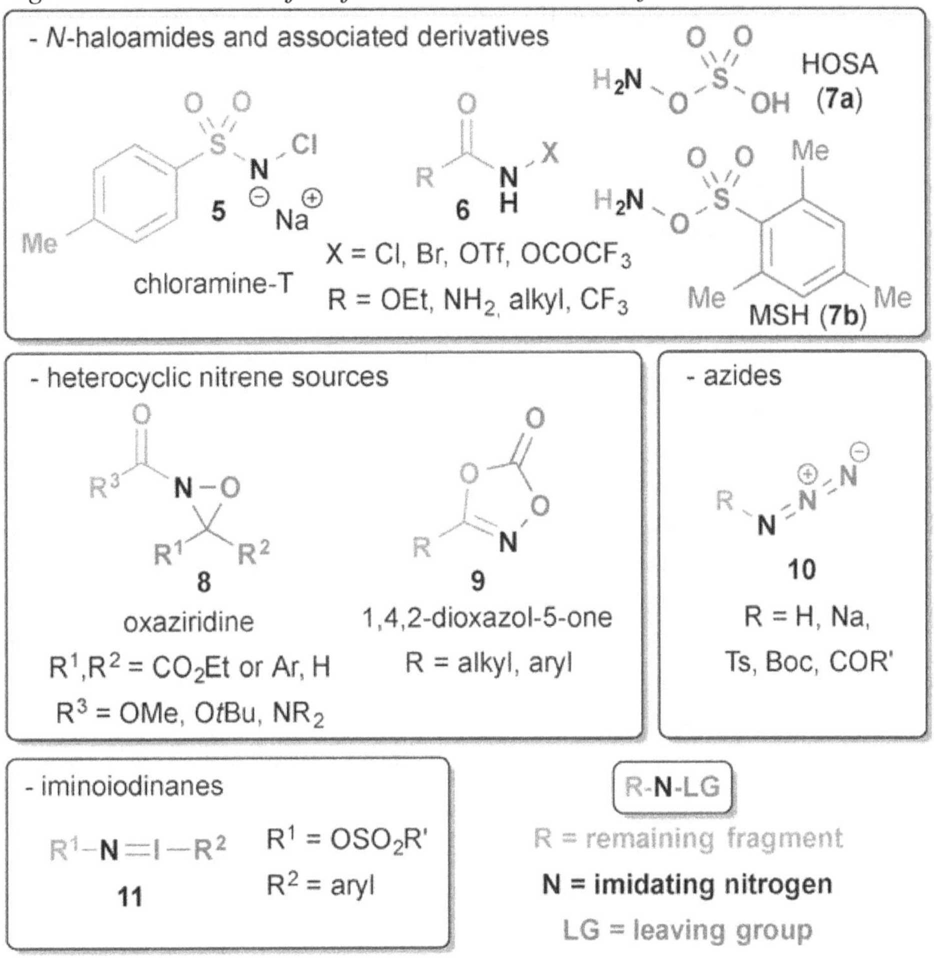

Imidation of Sulfides with N-OSO2R Carbamates

Reacting sulfides with N [(methansulfonyl)oxy]carbamate as imidating agent was highly efficient, leading to the corresponding N-(ethoxycarbonyl) sulfimides in good yields independent of the nature of R1 and R2. Recently, this chemistry was extended by the development of a diastereoselective version of this reaction applying mesyloxycarbamate (S)-23 as nitrogen source in combination with a chiral rhodium Catalyst. The use of additives allowed to increase the diastereoselectivities in the formations, which could finally be isolated with a d.e. of up to 90% in good to high yields. Results from mechanistic and crystallographic investigations suggested the involvement of a Rh(II)–Rh(III) complex and the formation of a metal-nitrene intermediate, (Khiar, 1993).

Imidation of Sulfides with N,O-bis(Trifluoroacetyl)-Hydroxylamine

N,O-Bis(trifluoroacetyl)hydroxylamine is an inexpensive imidation agent obtained by reaction of hydroxylamine hydrochloride and trifluoroacetic anhydride (TFAA). However, it showed low reactivity in sulfide imidation under many reaction conditions. Interestingly, the corresponding lithium salt showed a higher efficiency along with an increased stability, allowing bench storage of this salt at room temperature under air. Finally, combining of 6f with a catalytic amount of copper salt in dimethoxyethane (DME) led to sulfide imidations providing N-trifluoroacetyl sulfimidesin yields ranging from 65% to 95%. Next, an enantioselective synthesis was demonstrated following an entirely different approach. The asymmetric sulfide imidation was performed with chiral nitridomanganese(V) complex 26 that was activated by trifluoroacetic Anhydride. The reaction proceeded smoothly leading to products with moderate to very good enantioselectivities in high yields. Despite the significant efficiency of this procedure, the chiral manganese complex had to be applied in a stoichiometric amount, which limited the synthetic applicability, (Kunieda, 1976).

Imidation of Sulfides by Halogenation/Imitation

Even if some of the protocols described in the previous chapters implied a nucleophilic substitution in the reaction mechanism, all processes had in common that they involved a unique imidating reagent (RN-LG). Here, the imidation strategy is different as it relies on the use of at least two reagents performing an electrophilic halogenation/imidation sequence. These methods are presented here due to their mechanistic similarities. tert-Butyl hypochlorite (t-BuOCl) is a well-known reagent used for electrophilic chlorination. In the presence of sulfides at low temperature, t-BuOCl yields tetravalent intermediate which can react with amide anions to form the corresponding sulfimides. in moderate to good yields. The major drawback of this method lies in the choice of the amide anion which has to be acidic enough to avoid a competitive re-protonation by the generated t-BuOH. Consequently, aniline and methylamine cannot be used as amide sources in this reaction. Following the same strategy with a combination of t-BuOCl and a chiral alcohol (L-menthol) allowed obtaining optically active sulfimides, albeit with only low to moderate enantioselectivities. Presumably, the reaction proceeded via L-menthyloxysulfonium chloride, which reacted with the deprotonated amides to give diarylsulfimides. The low asymmetric induction (up to 47%) was suggested to be a consequence of an insufficient control in the diastereoselective formation as revealed by the isolation of the perchlorate anion equivalent and its structural analysis by NMR spectroscopy. Finally, a mechanistic switch (inversion versus retention) in the conversion was discovered, which depended on the acidity of the amidate anion. Thus, ordinary amidate anions such as N-(chloroacetyl) and N-tosyl amidates substitute the L-menthyloxy group with inversion of the configuration while the more acidic N-(dichloroacetyl), N-(trichloroacetyl) and N-(trifluoroacetyl) imidates undergo substitution with retention of the configuration, (Fernandez., 2003).

Figure 4. Sulfonium-aided coupling of aromatic rings

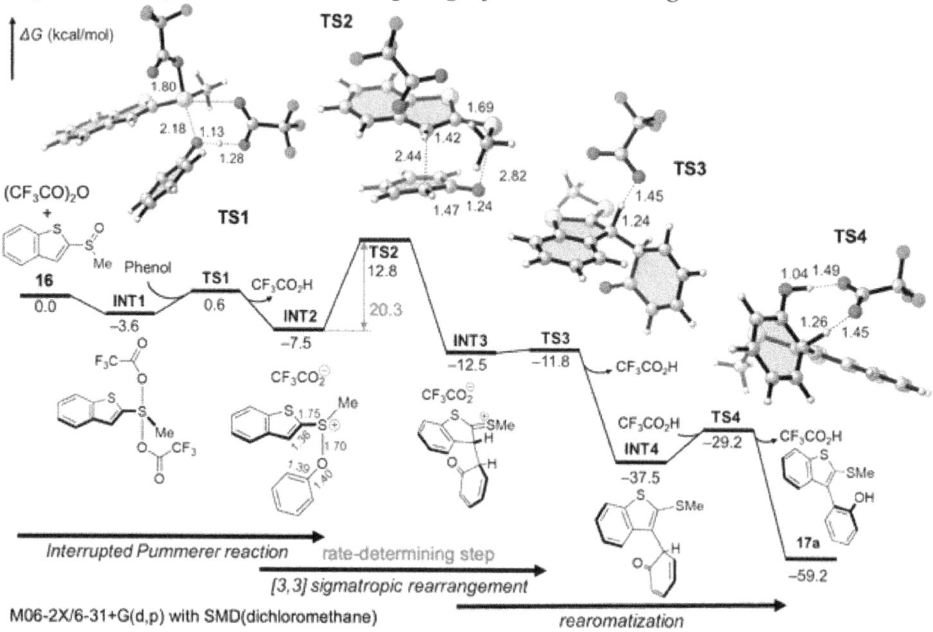

Sulfide Activation by Electrophilic Halogenation

The sulfide activation by electrophilic halogenation can also be achieved by using other Hal+ sources. Iodine and N-bromosuccinimide (NBS) were applied in synthesising N-cyano sulfimides starting from sulfides and cyanogen amine. Thus, with a combination of molecular iodine and sodium tert-butoxide as base, sulfimides 31 were obtained from S-aryl and S-benzylsubstituted S-methyl sulfides in moderate to high yields. Using cyanogen amine with NBS and potassium tert-butoxide for the imidation led to higher yields, but the crude product mixture also contained significant amounts of sulfoxide. The imidation of diaryl and dialkyl sulfides could only be achieved with the NBS/t-BuOK combination, while almost no reaction occurred with the I2/t-BuONa system. In a recent industrial study, a combination of 1,3-dibromo- 5,5-dimethylhydantoin (as brominating agent), 2,2,2-trifluoroacetamide (as nitrogen source) and sodium hydride (as base) was applied affording the corresponding N-trifluoroacetyl sulfimides in yields ranging from 71 to 88%.Recently, it has been demonstrated that alkyl amine–Br2 mixtures allow sulfide imidations providing intermediates that can directly be converted to N-alkyl sulfoximines. That is of particular interest for the synthesis of N-methylated sulfoximine derivatives (Banthorpe, 1964).

Figure 5. Imidations of sulfides with N-chloro sulfonamides

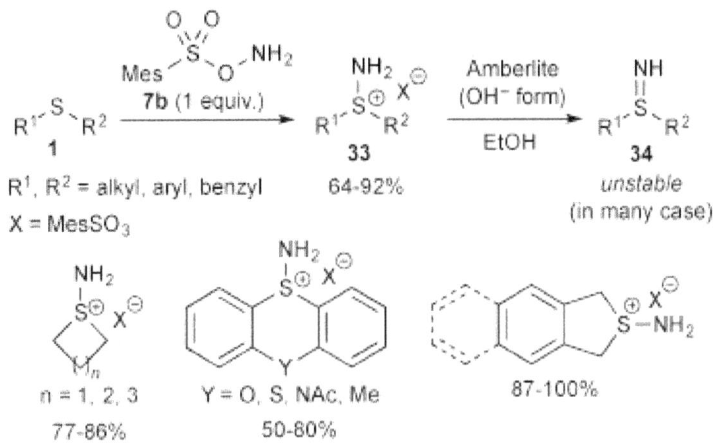

Imidations with MSH and Related Reagents

Sulfur imidations with hydroxylamine-O-sulfonic acid (HOSA) have been known since the early 1960 but only with the introduction of O-mesitylensulfonylhydroxylamine (MSH), this sulfur imidation strategy became popular. MSH is a powerful aminating reagent able to react with many nucleophiles. In general, an equimolar mixture of sulfides 1 and 7b in dichloromethane leads to the corresponding crystalline S-aminosulfonium salts in moderate to excellent yields. A subsequent basic workup transforms these salts into sulfimides. The free NH sulfimides 34 can also be obtained by passing an ethanol solution through a column filled with amberlite (OH_ form) ion-exchange resin. MSH can be applied for imidation of dialkyl, diaryl, alkyl-aryl or cyclic sulfides providing the corresponding sulfoniumsalts in good to high yields. The stability of the products is highly substrate dependent. The reactivity of MSH in reactions with diphenylsulfide and diphenylsulfoxide was compared to those of similar Compounds containing a H2N–OR site (with variations of the OR leaving group). The results showed a slight superiority of MSH over its analogues in the synthesis of sulfiliminium salt. When sulfoximinium salts were targeted, however, MSH was the only reagent that provided the products in good yields. Interestingly, the reaction of enantiopure sulfoxides with MSH yields the corresponding enantiopure sulfoximines with retention of the configuration at the sulfur stereogenic centre. When working withMSH, strict safety rules have to be followed as the reagent is thermally unstable and explosive. It should be stored below 0 1C and/or in dichloromethane solution. Safety concerns have also led to a major alternation of the original MSH synthesis resulting in an industry-devised protocol being applicable on a kg scale. The N-methyl version of MSH (Me-MSH) has recently been used for the direct synthesis of N-methyl sulfoximines via the corresponding N-methyl sulfilimides (Cox, 1963).

BACKGROUND OF CURRENT CONTEXT OF RESEARCH

Starting from an optically active sulfoxide, the imidation with Me-MSH was demonstrated to be stereospecific. Sulfur has the electronic configuration. Sulfur exists in acyclic and cyclic Sn species or allotropic form owing to have a marked tendency to react with itself (catenation). The sulfur-sulfur bond in elemental sulfur is probably a resonance hybrid for which S8 ring exists in the energetically favoured crown shape. Sulfoxides are either colorless liquids or solids with low melting points. For instance, dimethyl sulfoxide is one of the common amphiphatic solvents of low toxicity. It is used as a solvent in a wide variety of industrial processes, such as manufacture of polyacrylonitrile fibers, pesticides, in the extraction of aromatic hydrocarbons from refinery streams and for industrial cleaning. It is also used as a solvent for in vivo administration of water insoluble drugs and antitoxins. Omeprazole is a proton pump inhibitor used to treat acid-induced inflammation and ulcers of stomach and duodenum. A number of other acid secretion inhibitors such as Rabeprazole, Lansoprazole, Pantoprazole, have also been developed. Other biologically active compounds containing sulfonyl moiety include Modafinil and Sulindac which have been used in the treatment of narcolepsy and inflammation respectively. A few compounds containing sulfoxide as the structural motif serve as important intermediates in the manufacture of certain veterinary drugs and a few others are engaged in lubricants industry. They are also engaged in the large-scale separation of radioactive and less common metals. Alike sulfoxides, sulfones also play a prominent role in the field of organic synthesis, pharmaceuticals. They constitute a key skeletal framework of many natural products as well as pharmaceuticals and have attracted considerable attention because of their important biological properties. Particularly, aryl sulfones show various pharmacological properties such as anti-tumor, anti-inflammatory, and anti-fungal activities, including inhibition of HIV-1 reverse transcriptase. Many compounds containing sulfonyl group are known to exhibit antileprotic, antitumor, antiproliferative, antiresorptive and anticancer properties. Synthesis of sulfoxides or sulfone has attracted a great deal of attention from researchers. Numerous protocols have been reported describing a variety of methodologies used to synthesize sulfoxides and sulfones. This review accounts for the concise report of various protocols regarding synthesis of sulfoxides and sulfones. It also focuses on oxidative transformation of sulfides to corresponding sulfoxides and sulfones (Sheradsky, 1971).

Methodology of Current Context of Research: Synthesis of Sulfoxides

Synthesis of sulfoxides can be achieved by employing two different routes. (A) Non-oxidative routes (B) Oxidative routes

Non-Oxidative Routes

The most attractive pathway for the synthesis of aryl ketones involves Lewis acid catalysed acylation of arenes with acid chloride. Similarly, Lewis acid catalyzed sulfonylation of arenes with thionyl chloride is a well-known path for the synthesis of diaryl sulfoxides. Chasar and Pratt extended this protocol for the synthesis of ortho substituted diaryl sulfoxides by the reaction of aryl sulfinyl chloride with substituted phenols. Both the aforementioned Lewis acid catalysed protocols suffer from the drawbacks as regards the use of environmentally non-benign anhydrous aluminium chloride as the catalyst. In circumventing this problem, a Lewis acid-free protocol was proposed by Strecker. In which, the synthesis of dibenzyl sulfoxides was achieved by using methyl magnesium halide and thionyl chloride. The most popular non-oxidative approach for synthesis of the sulfoxides involves sulfinate methodology, wherein sulfoxides can be obtained by nucleophilic displacement on sulfinyl derivatives like sulphites, sulfinamides, sulfinyl chloride or sulfinic esters with organometallic reagents. Strecker reported for the first time that, the reaction of diethyl sulphite with two equivalents of Grignard's reagent in ether medium is a useful route for the synthesis of symmetrical sulfoxides. On the other hand, Montanari, Mislow and Wudl independently reported the synthesis of optically active sulfoxides using sulfinamides as precursors. Monteiro *et al.* synthesized a series of racemic β-keto sulfoxides by the reaction of arenesulfenate esters (methyl benzenesulfinate) with α-methylene carbonyl compounds in presence of sodium hydride. In an approach towards the synthesis of unsymmetrical sulfoxides, 2-trimethylsilylethyl benzene sulfinate was used as a precursor. It was prepared by the reaction between sodium benzene sulfinate and 2-trimethylsilylethyl chloride and it was then allowed to react with alkyl or aryl halide in the presence of potassium fluoride–calcium fluoride mixture as the catalyst. In this process, alkylation at sulfur with concomitant loss of trimethylsilyl fluoride as well as ethene furnishes the desired unsymmetrical sulfoxide. In an interesting approach, unsymmetrical sulfoxides have been prepared by the reaction of organo lithium or organo magnesium compounds with DABSO {1,4-Diazabicyclo[2.2.2]octane bis(sulfur dioxide)}. The necessary DABSO was prepared by the reaction of DABCO with sulfur dioxide. In this protocol the formation of sulfoxide was proposed to proceed through the generation of trimethylsilyl sulfinate intermediate. Apart from the aforementioned protocols, a few more non-oxidative protocols are also available for synthesis of sulfoxides. However, most of them are associated with some limitations such as handling of lachrymatory reagents, expensive starting materials, and drastic reaction conditions. Hence, these methods are not preferred for the synthesis of the sulfoxides. On the other hand, chemoselective oxidation of sulfides to sulfoxides is the most preferred pathway in the synthesis of sulfoxides. In this chapter we have also relied on this approach. Thus, it would be quite relevant to take a brief account of the protocols reported earlier for chemoselective oxidation of sulfides to sulfoxides, (Li, 2013).

Oxidative Routes

Chemoselective and controlled oxidation of sulfur from organosulfur compounds to sulfoxides is a challenging research topic in synthetic organic chemistry. A useful number of methods are available for the oxidation of sulfides to sulfoxides. Traditionally, although strong oxidants, like nitric acid or KMnO4 can be used for sulfoxidation, in tune with current demands of green chemistry, an array of reagents and reagent-catalyst combinations have been introduced for this oxidative transformation. The methods or the reagents available for the oxidation of sulfides to sulfoxides can be broadly classified into following categories (De, 2013).

Chromic Acid

Chromic acid in glacial acetic acid is a promising reagent-reaction medium consisting of many oxidation reactions. However, due to the high tendency of sulfides to undergo over oxidation to sulfones, it is not conventionally used for the oxidation of sulfides. Chromic acid-pyridine combination has been demonstrated to be highly effective in chemoselective and controlled oxidation of sulfide to sulfoxides.

Nitric Acid and Oxides of Nitrogen

Nitric acid had been the reagent of choice in oxidation of sulfides to sulfoxides. However, the main drawback associated with the use of nitric acid is the possibility of concomitant nitration of the substrate or other side reactions. Few other oxides of nitrogen employed for the oxidation of sulfides are: acyl nitrates, nitroniumhexafluorophosphate, thallium nitrate, ceric ammonium nitrate, nitrogen tetroxide, polyvinylpyrrolidine supported dinitrogen tetroxide (PVP-N2O4) etc. (Gao, 2013).

Halogen, Hypohalites

Organic sulfides on reaction with molecular halogen are known to form halosulfonium salts, which can undergo hydrolysis at the presence of water to yield sulfoxides. Based upon this concept, halogens, especially bromine or *in situ* generated bromine have been employed in oxidation of sulfides. The process suffers from drawbacks as regards formation of sulfonic acid and sulfinic acid as the by-products and the possibility of halogenation of the aromatic ring. By using hypohalites as oxidant, lower dialkyl sulfides are reported to undergo oxidation directly to sulfones while higher sulfides get converted to sulfoxides. Sodium meta-periodate (NaOI4) in excess amount has also been reported for selective sulfoxidation. Using NaOCl as an oxidant, TEMPO-linked metalloporphirines have been shown to be efficient catalysts in selective oxidation of sulfides to sulfoxides. Iodosobenzene has emerged as an efficient oxidizing agent for the oxidation of unsaturated or otherwise sensitive sulfides to corresponding sulfoxides. Iodoxybenzoic acid (IBX)-tetraethyl ammonium bromide has also been shown to be efficient oxidant-catalyst combination in chemoselective oxidation of numerous sulfides. The method offers the advantages of short reaction times, functional group compatibility and non-formation of over oxidation products. Bravo and co-workers demonstrated the use of molecular bromine and H2O2 combination for sulfoxidation. The reaction is known to follow free-radical pathway to furnish the desired product. Cetyltrimethylammonium tribromide (CTMATB) prepared by the reaction between molecular bromine and cetyltrimethylammonium bromide has been demonstrated as an efficient reagent in controlled and chemoselective oxidation of sulfides to sulfoxides. In fact, CTMATB serves as a slow-release source of bromine. The reagent mainly offers the advantages of easy handling of the oxidant and controlled oxidation (Guo, 2019).

Peroxides, Peracids and Ozone

Peroxy compounds such as hydrogen peroxide, organic hydroperoxides, peracids and ozone are well known sources of active oxygen and have been used for the oxidation of sulfides to sulfoxides. Among organic hydroperoxides, urea-hydrogen peroxide (UHP), *tert*-butyl hydroperoxide

(TBHP), 2-hydroperoxyhexafluoro-2-propanol, flavin hydroperoxide and α-azohydroperoxide are well documented as oxidants for sulfoxidation. In 1928 Lewin reported the use of perbenzoic acid for sulfoxidation of sulfides at room temperature while Sakuraba *et al.* reported the use of peracetic acid for the same oxidation. On the other hand, *meta*-chloroperbenzoic acid, although shock sensitive and potentially explosive in nature, is the most preferred oxidant in the oxidation of sulfides to sulfoxides (Yuan, 2020).

Amongst several peroxy compounds, hydrogen peroxide is an economical and commercially available oxidizing agent. Water, being the only by-product in hydrogen peroxide mediated oxidation, has dragged more attention. It is worth noting that hydrogen peroxide is, although a good oxidant from environmental point of view, its activity is low and the same can be increased by using appropriate reaction medium, catalyst or temperature. The efficacy of H2O2 in combination with number of catalysts or using different solvents has been reviewed (Zhang, 2020).

For instance, Molybdenum salt, immobilized vanadyl, cobalt or nickel alkyl phosphonates, zinc bromide, Ga(OTf), glacial acetic acid, poly(N-vinylpyrrolidone) (PVP) and poly(4-vinylpyridine), vanadium complex [VO2F(dmpz)2] (dmpz = 3, 5-dimethyl pyrazole) have been reported for the oxidation of sulfides, (Yanagi, 2021).

Rostami *et al.* and Zolfigol *et al.* described the use of boric acid and silica vanadic acid as highly efficient Bronsted acid catalysts in chemoselective oxidation of sulfides using hydrogen peroxide as oxidant under solvent-free conditions. Recently, Shen *et al.* reported on the influence of pH in Bronsted acid catalyzed sulfoxidation of sulfides using H2O2 as an oxidant. He has concluded that the performance of the catalyst in this oxidation does not depend upon the amount of the catalyst but it depends upon the pH of the reaction medium. With the choice of any Bronsted acid as catalyst at pH 1.30, more than 98% conversion with high selectivity and excellent substrate tolerance has been reported (Bur, 2004).

Attempts have also been made to check the efficacy of hydrogen peroxide in the oxidation of sulfides in various reaction media. In early studies acetone and acetic acid were the preferred solvents for the oxidation of sulfides using H2O2 as an oxidant. In an interesting study, Drabowicz *et al.* reported that the rate of sulfoxidation can be increased substantially by changing the reaction medium from acetone to methanol while Zu and co-workers have reported the use of phenol as a solvent in expeditious sulfoxidation of sulfides using 30% H_2O_2 as an oxidant. Oxone is another commercially available oxidant that has attracted considerable attention these days. Chemically it is a 2:1:1 ternary composite of $KHSO_5$, K_2SO_4 and $KHSO_4$. The only drawback with the use of Oxone is its solubility in water. This poses problems in using the same in oxidative transformations carried out in organic solvents. Despite this limitation, the use of Oxone in various organic transformations is well documented. At the same time, it is worth noting that reports on its use in the oxidation of sulfides are scanty. Greenhalgh was the first to report the use of Oxone-wet alumina combination to bring out oxidation of sulfides to sulfoxides in CH_2Cl_2 medium under reflux condition. Kroop *et al.* have reported the use of silica gel-Oxone combination to effect sulfoxidation. In a recent communication, Yu and his colleagues demonstrated that selective oxidation of sulfides to sulfoxides can be carried out by using Oxone alone as an oxidant in ethanol medium. However, very long reaction times (12 h) and reflux condition are the two main drawbacks associated with this protocol (Feldman, 2006).

Synthesis of Sulfones

Alike sulfoxides, synthesis of sulfones can be achieved following non-oxidative as well as oxidative routes.

Non-Oxidative Routes

Sulfonylation of aromatic compounds with arenesulfonyl halides or arenesulfonic acids has been the most preferred non-oxidative approach for the synthesis of sulfones. For instance, Bandgar and co-workers have reported the synthesis of sulfones by zinc catalysed sulfonylation of arenes using *p*-toluene sulfonyl chloride as the sulfenylating agent. The reaction conditions involve thermal as well as microwave activation. Recently, the marketed antiulcer drug Zolimidine and other aryl sulfones have been synthesized by CuI assisted coupling of sulfinate and aryl halide at the presence of D-glucosamine as a green ligand. Friedel-Craft's sulfonylation method for the synthesis of diaryl sulfones using ionic liquid was introduced by Salunkhe. In this protocol, the ionic liquid used viz. [bmim] Cl-AlCl3, has been reported to serve as the reaction medium as well as catalyst, (Akai, 2007).

For regioselective sulfonylation of heterocycles with aryl sulfonyl chlorides, activation of C–H bond using metal-based catalysts has been demonstrated to be essential. Saidi *et al.* have developed a method for activation of *meta* C-H bond in 2-phenylpyridine using Ruthenium complex. Thus, selective sulfonylation of 2-phenylpyridine to afford *meta* phenylsulfonyl -2-phenyl pyridine was achieved. Pan *et al.* recently reported an electrochemical synthesis of β-sulfonylesters using of aryl sulfinates and β-keto ester compounds as substrates and ammonium iodide as an electrolyte. Regioselective ring opening of epoxide through addition of salt of sulfinates is another attractive strategy for the synthesis of functionally diverse sulfones. Chumachenko and Sampson have reported the synthesis of aryl sulfones by ring opening of ethylene and propylene oxide with zinc sulfinate in aqueous medium. Murthy *et al.* have recently reported a green and catalyst-free protocol for the synthesis of β-hydroxy arylsulfones by heating together a mixture of sodium salt of aryl sulfinic acid and epoxide in water medium. Resultant of β-hydroxy arylsulfones serve as starting materials in the synthesis of vinyl sulfones as well as β-ketosulfones which serve as important intermediates in organic synthesis, (El-Sayed, 2011).

Chawla *et al.* developed a simple protocol for the synthesis of aryl vinyl sulfones by heating together a mixture of sodium benzenesulfinate and terminal epoxide in presence of lithium bromide as the catalyst. The reaction passes through the formation of β-hydroxy arylsulfones as the intermediate but, subsequent elimination of water to furnish corresponding vinyl sulfone is not regioselective. Thus, the resultant vinyl sulfone is a mixture of terminal and internal vinyl sulfone. Phenylsulfonyl acetonitrile is an active methylene compound and serves as a valuable synthon in organic synthesis. It could be prepared in excellent yield by heating together a mixture of sodium benzene sulfinate and 2-chloroacetonitrile in aqueous-dimethylformamide.

Oxidative Routes

Oxidation of thioethers directly to sulfones is another, simple and most useful route for the synthesis of sulfones. A literature survey revealed that a variety of reagents under the range of experimental conditions have been reported for the direct oxidation of sulfides to sulfones. It is worth noting that regardless of the medicinal importance of sulfones, much attention has earlier been paid towards chemoselective oxidation of sulfides to sulfoxides and the protocols developed for the same have been extended towards the synthesis of sulfones. Nitric acid, KMnO4, bromine, NaOCl, ozone, peracids, *t*-BuOOH are few of the oxidants used in early days. Few reports are also available on the use of magnesium monoperoxyphthalate (MMPP) and polyvalent and hypervalent iodine compounds. In addition to these, Jereb *et al.*reported a catalyst-free, solvent-free and functional group compatible protocol for the oxidation of sulfides to sulfones using 30% H2O2 as the oxidant. The main drawback of the developed protocol involves the necessity of elevated temperature (75 0C) and longer time for complete oxidation. Recently, task specific bifunctional ionic liquid viz. [pmim]IO4 has been reported for selective oxidation of sulfides to sulfones. In this reaction, ionic liquid serves as a solvent as well as an oxidant. Oxone has also been the choice of a few researchers to bring out oxidation of sulfides to sulfones. However, the developed protocols suffer from the drawbacks like necessity of excess equivalent of Oxone, or inorganic supports like montmorillonite, alumina etc. coupled with thermal / MW activation. The problem of using excess Oxone was circumvented by Kupwade *et al.* They have reported diethylamine catalysed eco-benign protocol for controlled, selective as well as scalable protocol for oxidation of sulfides to sulfones using Oxone. This report also mentions uncatalysed oxidation of sulfides to sulfoxides, (García-Galán, 2008).

RESULTS OF CURRENT CONTEXT OF RESEARCH

This review spotlights sulfonamides from different sides, including history, structure-activity relationship, chemistry, methods of classification and up to date ways for their synthesis. Moreover, the review includes a discussion of the varied pharmacological effects of sulfonamides. Sulfonamides have a good range of pharmacological activities like oral diuretics (furosemide, indapamide, chlorthalidone, thiazides); anticancer (E7070), carbonic anhydrase (CA) inhibitors (CAIs) (acetazolamide, dichlorophenamide, dorzolamide and brinzolamide), antiepileptics (zonisamide and sulthiame), for rheumatoid arthritis (Sulfasalazine) antiviral (Darunavir) anti-inflammatory (celecoxib), antibacterial (Sulfadizine), as ophthamologicals (Dorzolamide) anticonvulsant (Zonisamide) cycloxygenase 2 (COX2) inhibitors (valdecoxib). These days, novel pills are launched, like apricoxib and pazopanib, which additionally include this organization. The sulfonamides competitively inhibit vitamin Bc synthesis in micro-organisms and subsequently inhibit multiplication of bacteria but donot actively kill them. They need to be used against most gram-positive and lots of gram-negative bacteria, also as some fungi Synthesis of sulfonamide using primary amines and aryl sulfonyl chloride Youn *et al.* reported the preparation of sulfonamide using aryl primary amine and aryl sulfonyl chloride employing pyridine as a base at 0-25 °C. They've discovered 100% yield when aniline is used as a primary amine and benzene sulfonyl chloride or 4-nitrobenzyl sulfonyl chloride as sulfonylation agent, (Elgemeie, 2019).

Quantitative yield additionally stated for the reaction between p-toluidine and tosylchloride. The purpose of the take a look at changed into a regioselective synthesis of 3- Arylindoles from N-Ts-Anilines and styrenes. Rattanaburi et al. reported Fe3O4-DIPA catalyzed sulfonamide preparation with excellent yield (98%) where the reactant and Fe3O4-DIPA in dichloromethane (DCM) were shaken at room temperature (RT) for reaction completion. The catalyst became separated with the aid of magnetic separation and reused. Chemo selective solvent free synthesis of sulfonamide using zinc oxide-nanoparticle was prepared with 95% yield. Synthesis started with primary amines, sulfonylation accompanied by way of acylation produced N-acylsulfonamides. Reusability of environmental-friendly catalysts turned into the principal advantage of the protocol. Also, the invention of surprisingly efficient catalyst CsF-Celite for sulfonylation reaction was reported by Chemoselective solvent free neat reaction has been demonstrated for the preparation of various sulfonamides with high yield. Over the decades, researchers suggested use of neat pyridine or its combination with polar solvents for the synthesis of sulfonamides of primary amines e.g., filed a patent on ramoplanin derivatives having antibacterial activity in 2006 where they have reported the synthesis of N-phenylbenzene sulfonamide with 90% yield. Neat pyridine changed into delivered to amine substrate at 0oC and the reaction became performed at room temperature (RT) after addition of sulfonyl chloride. Kato et al. suggested synthesis of a chiral aromatic sulfonamide that are in addition passed through spontaneous fast resolution to provide chiral crystal. Polymer supported pyridine became used as a base and dichloromethane as a solvent to afford 92% yield. Alba et al. synthesis a monosulfonamide and bis-sulfonamide with 90% yield where sulfonamide prepared using pyridine as a base in THF-solvent. The synthesized sulfonamides were used as an organo-catalyst, i.e. hydrogen-bond donors in organo-catalyzed ROP of locations. Moderate to suitable yield was stated in numerous articles for the sulfonylation of primary amines using another commercially considerable organic base triethylamine (TEA). Tosylation was carried out by TEA as a base in THF solvent 86% yield was obtained by Kurkin et al. Synthesis procedure was simple; TEA added drop wise to the solution of aniline in THF, the mixture was stirred in ice bath. Benzenesulfonyl chloride was added drop wise and stirred at RT for 6 h for reaction completion (Scozzafava, 2003).

Conway and Gribble reported 85% yield of N-phenylbenzene sulfonamide by reaction between aniline and benzene sulfonyl chloride in diethyl ether at 0 °C. Qiu et al. reported reaction between domino Aryan precursor and sulfonamide where sulfonamides were prepared with 85% yield using TEA as a base in DCM. The use of an inorganic base like potassium carbonate was also explored in sulfonylation reaction where the researchers achieved up to 78% yield whilst reactions have been done in PEG-400 solvent. The evolved protocol turned into handy because of heterogeneous response mass in which the bottom may be easily separated from the reaction mass through filtration. Recovery and reuse of PEG-400 is another advantage of the reported method from economic and environmental aspects. Newcomer et al. reported poor yield (only 44%) when strong inorganic base like sodium hydroxide was employed for the sulfonamide synthesis reaction. In a magnetic stirrer acid chloride and amine were mixed collectively. The addition of 10% NaOH changed into performed in quantities and the response combination stirred for 1 h at RT to produce N-phenylbenzenesulfonamide. The main intention of the study was to prepare unalkylated benzene sulfonanilides to check rearrangement of the same to sulfones. Sulfonamide synthesis by reaction between aqueous solutions of primary amine and sulfonyl chloride at RT using sodium carbonate as a base was reported the research group prepared a series of sulfonamide- 4-substituted-1, 2, 3-trizolyl nucleosides and evaluated their activity against tumor cell lines RCC4 and MDA-MB-231 (Harter, 2004).

Synthesis of Sulfonamide Using Amines and Thiol

Another method for the synthesis of sulfonamide of primary amines has been explored by way of the researchers where SH group of aromatic thiol was insitu oxidized through the use of oxidizing agent and chlorinated before the amide bond formation. Bahrami et al. reported stage wise reaction between aniline and benzene thiol using hydrogen peroxide as an oxidizing and zirconium chloride as a chlorinating agent. In degree-2, pyridine was used as a base and the typical 98% yield changed into completed. They have claimed environmentally green and economical protocol where the novel H2O2-ZrCl4 reagent system offered fast reaction at room temperature to furnish excellent yield (92-98%). The same studies also mentioned a stage wise process a combination of hydrogen peroxide and thionyl chloride reagent system were used for direct oxidative chlorination of various thiol derivatives to the corresponding sulfonyl chloride intermediates, which had been in addition converted to sulfonamide in level 2 via reacting with one of a kind amines using pyridine as a base in acetonitrile or water solvent to afford 94- 98% normal yield. Thiol may be transformed to sulfonyl chloride in many methods. Aside from the oxidative chlorination of thiol to sulfonyl chloride, in-situ conversion of heteroaromatic thiols to sulfonyl choride the use of sodium hypochlorite as an oxidizing agent was reported to prepare sulfonyl chloride at low temperature, which was immediately trapped with benzyl amine to produce sulfonamide in almost quantitative yield. Scully and Bowdring prepared organic chloramines by mixing sodium hypochlorite and aq. The solution of a mine which has reacted rapidly with sodium arene sulfinate to form arene sulfonamide. Veisi et al. developed a novel method where an oxidizing chlorinating system was developed to synthesize sulfonyl chloride from various thiols by mixing NChlorosuccinamide and tertbutylammonium chloridewater system in acetonitrile solvent. The sulfonyl chloride is in situ transformed to the corresponding sulfonamide in a single pot. Bonk et al. reported novel system for controlling liberation of chlorine in combination of trichloroisocyanuric acid and benzyltrimethylammonium chloride in MeCN solvent. The sulfonyl chloride turned into generated, and in-situ reacted with diverse amines to diversify the scope of the protocol. The concept of green chemistry was applied for the first time in sulfonylation reaction wherestage wise reaction using iodine and ethanol was investigated. The reported procedure was novel and efficient since any metal, base, ligand or additive was not used in the reaction. The scope of the protocol was also very wide, which had covered the construction of a variety of primary, secondary and tertiary sulfonamides. Yu and Zhang mentioned for the first time that the use of benzene sulfonyl hydrazide for sulfonylation of various amines in presence of tertbutylhydroperoxide (TBHP) and ammonium iodide in water or acetonitrile (ACN) with moderate to good yield. TBHP was used as an oxidant and ammonium iodide (NH4I) as a catalyst to produce aryl sulfonyl hydrazide which is further reacted with amines to produce corresponding sulfonamides. A novel reagent 1-phenylsulfonylbenzotriazolewas synthesized for the conversion of various aliphatic and aromatic aminesand phenols to their corresponding benzenesulfonamides and benzenesulfonates (Stranix, 2006).

Sulfonamide synthesis using secondary amine slike Nmethyl aniline or diphenyl aniline was synthesized. The aim of the study was to explore the hemolytic cleavage and intermolecular radical-radical coupling reaction mechanism of 1,3- and 1,5-sulfonyl migration of N-arenesulfonyl phenothiazines and N-arenesulfonylphenoxazines. Excellent yield (97%) was achieved using pyridine as a base in chlorinated solvent by portion wise adding of benzenesulfonyl chloride. Terent'ev et al. reported an effective electrochemical synthesis of sulfonamides from arenesulfonohydrazidesor sodium p-methylbenzene sulfinate and amines. The reactions were carried out in an undivided cell using graphite anode and iron cathode at constant current density 35-40 mA cm-2 to efficiently produce fourteen various sulfonamides in 56-98%. Recently in 2018, various tertiary sulfonamide derivatives of pyridyl-indolebased heteroaryl chalcone in 60-90% yield was discovered. Reactions were carried out room temperature (RT) using weak inorganic base Na2CO3 in 50% THF:H2O solvent mixture. All the compounds were evaluated for carbonic anhydrase IX inhibitors and anticancer agents. Scheme 11; Synthesis of sulfonamide derivatives of pyridylindole based heteroaryl chalcone An open air, metal free oxidative coupling for the synthesis of sulfonamide mediated by phenyl trimethyl ammonium tribromide (PTAB). Initially benzene sulfinatereacted with PTAB to produce forms the corresponding sulfonyl bromide, then the nucleophilic reaction of amineswith sulfonyl bromide produced the desired products in 54-82% yield.

A ***diverse range of azoles and pyrazoles were converted to sulfonamides*** following simple and green procedure. Initially, aromatic sodium sulfinatewas reacted with N-Iodo or Nbromosuccinimide (NBS) to produce sulfonyl bromide or iodide, which was finally converted to sulfonamides by means of nucleophilic assault of azoles or benzimidazoles. Scheme 13; Synthesis of sulfonamides using substituted azole or benzimidazole and sodium sulfinates d- Synthesis of sulfonamide using heteroaryl amine Reaction between 2-aminopyridineand benzene sulfonyl chloride was reported by using organic base pyridine in 63% yield. Comparable low yield (29%) obtained when the same sort of reactions were carried out applying microwave irradiation (Anderson, 1992).

Iodine catalyzed reaction between heteroaryl amine and the sodium salt of benzenesulfonate was reported in 52% yield. Recently in 2017, synthesized various quinazoline sulfonamide derivatives and evaluated them as an anticancer agent against SKOV3, DU145, THP1, U937and COLO205 cell lines. Few compounds had been found to be energetic against THP1 and U937 cell lines. Different aryl, heteroaryl, alkyl and cyclopropyl sulfonyl chlorides were reacted with 2- chloro-6,7-dimethoxyquinazolin-4-aminein DMF and THF solvent mixture using sodium hydride as a base to produce the corresponding sulfonamide in 72-96% yield.

Synthesis of substituted quinazoline sulfonamide of secondary amines starting from sulfonic acid Microwave assisted preparation of sulfonamide starting from sulfonic acid was explored by in high yield. The reaction proceeded via sulfonyl chloride intermediate. 2, 4, 6-trichloro-[1, 3, 5]-triazine was added to a solution of p-toluenesulfonic acid in acetone, followed by TEA. The reaction mass was irradiated at 80 °C for 20 min, thesulfonyl chloride is isolated by filtration and further reacted with an allylamine to produce corresponding sulfonamide in 95% yield. Rad et al. reported one pot synthesis of sulfonamide from primary or secondaryamine by preparing amine sulfonate saltusing cyanuric chloride. Using a classical heating method with excellent yield. The reaction mechanism has also been explained by the research group. The reaction proceeds via a SNAr - type reaction between sulfonate anion and cyanuric chloride. The chloride ion attacks the sulfur atom to produce the corresponding sulfonyl chloride. In the next step, the amine liberates from ammonium in the presence of TEA, which reacts with sulfonyl chloride to afford sulfonamide products.

Preparation of secondary and tertiary sulfonamides using a metal catalyst Pd catalyzed sulfonamidation of aryl nonafluorobutanesulfonates has been investigated. The optimal reaction condition was established for N-arylation where combinations of Tris(dibenzylideneacetone)dipalladium [Pd2(dba)3] and binary phosphine ligand, t-BuXPhOS were used as highly active catalyst and K3PO4 in tert-amyl alcohol (TAA) was found as an optimal base solvent combination for the reaction. Though the instability of di-substituted aryl nonaflates is a major limitation, the wide substrate scope is the advantage of the protocol. Scheme 20; Pd catalyzed synthesis of sulfonamide Pd-catalyzed cross coupling reaction between substituted aryl halide and methane sulfonamide (scheme 21) was studied with high yield. Sulfonamide synthesis using borrowing Hydrogen catalyst Solvent free microwave assisted N-alkylation of primary sulfonamide with alcohol to produce secondary sulfonamide in very good yield was reported. The Ru-complex was used as a catalyst in the reaction to borrow hydrogen from alcohol. The intermediate aldehyde reacted with an amine to generate imine, which is eventually reduced by the catalyst to produce sulfonamide of the secondary amine. Lam et al. reported Cu catalyzed N-arylation of benzenesulfonamide using aryl boronic acid in quantitative yield. Diversity of cross coupling between arylboronic acid with a wide range of primary sulfonamide using the catalytic copper system is the main advantage of the invented protocol. Catalytic Cu(OAc)2/TEMPO in the air and catalytic Cu(OAc)2/O2 system worked well for the majority of the substrate. Another example of Cu catalyzed N-arylation of sulfonamide was disclosed by where coupling between aryl bromideor iodide and sulfonamide was carried out in excellent yield. DMF used as a solvent, K3PO4 and amino acid as a base and ligand. Invention of less expensive and more environmentally benign Cu(I)/amino acid catalyst system for C-N cross-coupling reaction is the novelty of the protocol (Gleckman, 1979.)

DISCUSSION OF CURRENT CONTEXT OF RESEARCH

Woolven *et al.* synthesized a stable complex DABSO, which was prepared by reaction between DABCO and gaseous sulfur dioxide. The complex was used as a safer source of sulfer dioxide, which was combined with Grignard reagent to produce sulfinate. Sulfuryl chloride was used to convert sulfinates to sulfonyl chloride, which was in-situ reacted with various amines to produce corresponding sulfonamides. Synthesis of sulfonamide using DABSO-Bis (sulfurdioxide complex) Indium catalyzed sulfonylation of sterically hindered and less nucleophilic amines were reported by in excellent yield. A wide variety of sulfonamides was synthized by using the new methodology. The research group claimed reusability of the catalyst up to five times. They proposed that, eletrophilic species RSO2 + might be generated by reaction between sulfonyl chloride and indium metal, which is further reacted with aminesto furnish sulfonylated product and active indium metal.

Synthesis of Sulfonamide Using Catalytic Indium

Cu catalyzed oxidative coupling between various amines and sodium sulfonates were investigated. Good yield and excellent chemo selectivity were observed since the transformation happened via a single electron transfer pathway. Synthesis of sulfonamide via Grignard discovered an innovative method for the preparation of sulfonyl chloride followed by sulfonamide. Aryl and heteroaryl bromides had been first converted into the corresponding Grignard reagents the usage of isopropylmagnesium chloride, which had been subsequently reacted with sulfuryl chloride and amines to produce sulfonamides. Recently a two-step sequential metal free process for the synthesis of sulfonamides using N,N disubstituted formamides as an amine and sodium sulfinates as sulfone source. Initially, formamide was decarbonylated by Potassium tert-butylate (KOt-Bu) to produce the corresponding amine in acetonitrile solvent. At the same time sodium sulfinatereacted with N-Iodosuccinamide (NIS) to afford sulfonyl iodide which eventually generated sulfonyl radical. Finally, nucleophilic attack of amine to sulfonyl radical provided desired sulfonamide in 24-91% yield, (Bushby, 1968).

Medicinal Chemistry Applications

Antimicrobial Sulfonamide compounds are a big class of synthetic bacteriostatic antibiotics which are still used today for the therapy of bacterial infections and other infections caused by different microorganisms. Additionally, they're referred to as sulfa drugs and were the most source of therapy against bacterial infections, before the introduction of penicillin in 1941. Additionally, the primary sulfonamide section is found in many clinically used drugs like diuretics, carbonic anhydrase inhibitors and antiepileptics. Sulfonamides are antimicrobial drugs with a broad spectrum of action, effective against Gram-positive and certain Gram-negative bacteria, like intestinal species. Sulfonamides show a good activity against E. coli, moderate towards Proteus mirabilis and Enterobacter species and weak towards Klebsiella, however they display no inhibitory activity in opposition to pseudomonas aeruginosa and Serratia species. They are effective against the species of Chlamydia genus. Sulfonamides also are powerful towards fungi (Pneumocystis carinii) and protozoa (Toxoplasma gondii). Sulfonamides differ in potency, but not in the spectrum of antimicrobial activity. Because of the lack of ability of bacteria to gather dihydrofolic acid from their environment, as part of microorganism's DNA biosynthesis, inhibition of dihydrofolic acid synthesis poses a desirable target for bacteriostatic agents. Inhibition of these enzymes has been achieved with early sulfonamides such as sulfanilamide. The formation of dihydrofolic acid is initiated with the aid of coupling pteridine diphosphate with p-aminobenzoic acid, which can then go through amid coupling with glutamic acid to shape dihydrofolic acid. Sulfanilamide displays comparable core structure to that of p-aminobenzoic acid and act as a competitive inhibitor. The Dihydrofolic acid formation is interrupted during the second step; because of a lack of acidic terminal to be had to couple with glutamic acid, hence dihydrofolic acid formation is interrupted. Synthesized and studied

Antimicrobial Activity of some novel thiourea, hydrazine, bisthiourea andfused pyrimidine derivatives containing sulfonamide moieties. Fourteen compounds were screened in vitrofor their antimicrobial activities against four strains of bacteria staphylococcus aureus (NCTC-7447), Bacillus cereus (ATCC-14579), Serratia marcesens (IMRU-70), Proteus mirabilis (NCTC-289) and two strains of the fungi Aspergillus ochraceus Wilhelm (AUCC-230) and Penicilliumchrysogenum Thom (AUCC-530) by the agar diffusion technique. some derivatives containing sulfonamido moieties New aryldisulfonamides were synthesized showing antibacterial activities against Gram positive bacteria (S.aureus ATCC 25953), (B.aureus ATCC 6633) and Gram negative bacteria (E.coli ATCC 11230) were done by Alyaret *et al.* (Thomas, 2007).

A series of novel heterocyclic thioureas containing sulfonamide moiety has been synthesized through condensation of isothiocyanatobenzene sulfonamide with a variety of heterocyclic amines. The newly synthesized compounds had been evaluated for their in vitro antibacterial activity against Streptococcus pneumoniae and Bacillus subtilis as examples of Gram-positive bacteria and Pseudomonas aeruginosa and Escherichiacoli as examples of Gram negative bacteria. They were also evaluated for their in vitro antifungal capacity towards a consultant panel of fungal strains i.e. Aspergillusfumigatus, and Candida albicans. Interestingly, some compounds showed similar or better activity comparison with the reference drug against the microorganisms examined. In the next year Ghorab et al. described a new series of novel fluorinated thiourea derivatives carrying sulfonamide moieties. Fluorinated pyridine derivative showed the highest antimicrobial activity (with MIC values ranged from 1.95 to 15.63 μg/mL). Interestingly, thiadiazole derivative and coumarin derivative exhibited selective antibacterial activities against Gram positive bacteria. Fluorinated pyridine derivative was the most active against HepG2 with IC50 value of 4.8μg/mL Molecular docking was performed on the active site of MK-2 with good results. Recently, synthesized a series of 4-((4- ydroxynaphthalen-1-yl)diazenyl)benzenesulfonamides which have been prepared by subsequent diazotization of sulfonamide derivatives and coupling with 1-naphthol in alkaline medium. Cyclization of 4- ((4-hydroxynaphthalen-1-yl) diazenyl)benzene sulfonamides with cinnamic acidin the presence of a basic catalyst afforded the novel naphtho[1,2- b]furans.Also,4-((4-

hydroxynaphthalen-1-yl)diazenyl)benzenesulfonamide can be cyclized with α- cyanocinnamonitrilesto afford 2-amino-3-cyano-4- phenyl-4H-benzo[h]chromenes.4-(4-amino-3,5-dicyano-6-iminopyridazin-1(6H)-yl) benzenesulfon amides were obtained at room temperature by treatment of 2-amino-1,1,3-tricyanopropene with a diazonium salt of sulfonamide derivatives. The newly synthesized compounds tested displayed variable invitro antimicrobial activities under these screening conditions. Interestingly, the tested compounds exhibited significant antifungal activities against the filamentous fungus (Aspergillus fumigatus) and unicellular yeast (Candida albicans). In the same year they described the new series of sulfonamides drugs linked with thiourea-carbamate side chains as a drug delivery system was synthesized. The thiourea-carbamate hybrid was connected to para-amino group which act as a prodrug to enhance sulfonamides activity. The novel hybrids sulfonamide-carbamates were synthesized by treatment of N-substituted 4- isothiocyanatophenyl sulfonamides (which synthesized in high yield via thiophosgenation of sulfonamides at room temperature in acidic medium (HCl), according to literature procedure) with ethyl carbamate in dry 1,4-dioxan at reflux temperature in the presence of triethylamine. Some Compounds exhibited activity against tested bacteria, (Özbek, 2007).

Anti-Carbonic Anhydrase Activity

Carbonic anhydrases catalyze the interconversion of carbon dioxide and bicarbonate. The main function of carbonic anhydrases is to modulate physiological pH, respiration, CO2 transport, electrolyte excretion, regulation and homeostasis. Sixteen CA isozymes have been identified in humans, CA I to CA XV, with different sub-cellular localizations. The role of CAs in cancer can be explained thinking about the metabolic conditions required by a cancer growing cell, i.e., a cell developing with a higher rate of replication with respect to a normal one. Any such circumstance calls for an excessive flux of bicarbonate into the cellular itself to be able to offer substrate for the synthesis of either nutritionally crucial components N (nucleotides) or cellular structural components (membrane lipids). On the other hand, H+ ions produced by metabolism are pumped from the intracellular compartment into the interstitial one, subsequently flowing into the blood. Still, the level of H+ ions in the interstitial compartment is higher in tumors compared with normal tissue. Among the CAs, CA IX and XII show close association with tumors, therefore, believed that inhibition of CA IX/CA XII could be an attractive alternative option for anticancer therapy. Sulfonamides 97-100 have anti-carbonic anhydrase activity in nanomolar concentration. However, due to their low selectivity, these CA inhibitors show good inhibitory activity against several isozymes including CA I, CA II, CA IX, and to some extent, CA XII. Thiry *et al.* later synthesized another class of sulfonamides and carried out an investigation of their potency against CA I, CA II and CA IX. It was found that even though the compounds were ineffective against CA I they were very effective against both CA II and CA IX. Furthermore, a docking study of 33 within CA II and CA IX, by the same group, indicates that the binding pockets of the two isozymes (CA II and CA IX) are very similar; if this is the case then it will make it difficult to target one isozyme over the other. Supuran et al. most recently reported that, by introducing a pyridinium moiety to the CA II/CA IX inhibitor, tumor acidification can be reduced by specifically inhibiting CA IX. Were reported and display good inhibition of CA IX over CA II. Specific inhibitors for CAIX Acetazolamide (AZA, anti- glaucoma), Zonisamide (ZNS, anti-convulsant) and Topiramate (TPM, anticonvulsant). Are known for their ability to inhibit CAs, and their X-ray crystal structures have been determined, which show tight binding of the inhibitors to CA II, CA VA and CA VB. Interestingly, it has been noted that, during clinical studies, obese patients experienced dramatic weight loss as a side effect. Therefore inhibiting CA II, VA and VB can reduce the rate of lypogenesis, and in turn, can be used as anti-obesity drugs. Commercially available inhibitors for CAV Vulloet al. have synthesized a wide range of sulfonamides such as and for anti-CA V activity, and it was found that several aminobenzolamides display low-nanomolar potency compared to registered CA inhibitors (Budavari, 1989).

Figure 6. Structures of few medicinally privileged sulfones

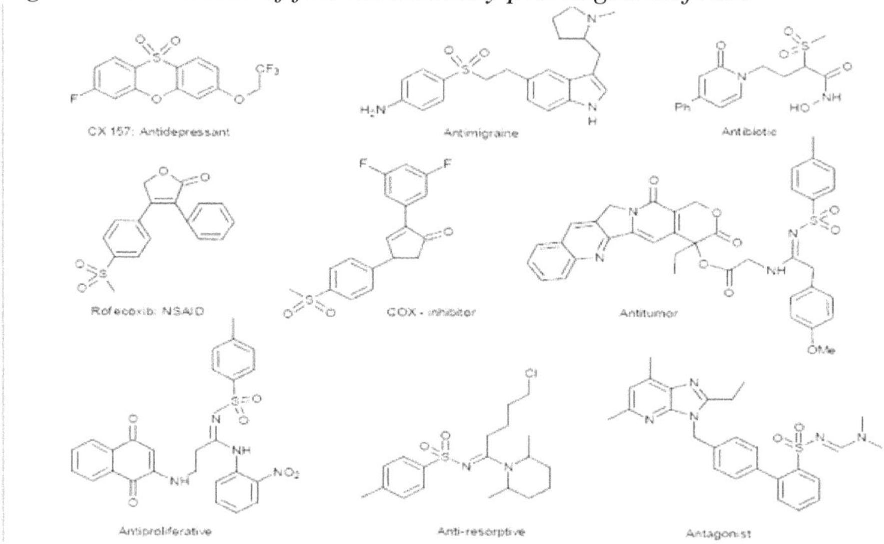

Antiviral HIV inhibitors Sulfonamides also possess activity against HIV proteases. HIV protease consists of a homodimer with aspartyl active sites (Asp and Asp), which have the ability to cleave difficult bonds, such as Tyr-Pro and Phe-Pro2. Thus far, several HIV protease inhibitors are clinically available, and are often used in conjunction with reverse transcriptase inhibitors to deliver the multi-drug treatment known as the Highly Active Anti-Retroviral Therapy. It was found that non-peptidic protease inhibitors display higher bioavailability as well as a slower excretion rate compared to the conventional peptide-base protease inhibitors. Among those protease inhibitors Amprenavir and Tipranavir aresulfonamide derived drugs. Most drugs achieved their high potency due to their ability to bind rigidly and optimally into an active site. However, high mutation rates within HIV viral particles render many anti-HIV drugs inactive, furthermore, they pose tremendous challenge to antiretroviral therapy. A more recent idea of overcoming this problem is to synthesize a drug that, instead of binding tightly to the protease active site, binds tightly with the protein backbone since the protein backbone are only minimally deformed during mutation (Cairns, 2008).

CONCLUSIONS AND FUTURE SCOPE

Due to the importance of sulfoxides and sulfones in organic chemistry, biology and polymers development of efficient protocols for the synthesis of sulfoxides and sulfones has been the subject of many research studies. Although the literature is enumerated with several approaches for the synthesis of sulfoxides and sulfones, the most rational and preferred pathway for their synthesis is the oxidation of sulfides. Several protocols have earlier been reported for these oxidative transformations. Among these protocols, oxidation using H2O2 are economical and benign, However, the methods developed suffer from the drawbacks as regards necessity of expensive catalyst, long reaction times, high temperatures, generation of unnecessary waste and primly, unsatisfactory performance in medium to large-scale synthesis. Recently Oxone has emerged as ecobenign alternative for oxidation reaction. It has proven its efficacy for selective and scalable oxidative conversion of sulfides to corresponding sulfoxide or sulfone. Very recently we have mentioned the use of Oxone for chemoselective, uncatalysed oxidation sulfides to sulfoxide as well as diethylamine catalysed oxidation of sulfides to sulfone. It is most simple, energy efficient and scalable protocol available for the oxidation of sulfides, complying with several key requirements of green chemistry.

In this context, the development of a late-stage imidation leading to NH- or specifically N-functionalised products would be desirable. From a stereochemistry point of view, several methods are available for the direct synthesis of optically active sulfimides, which can then be oxidized stereospecifically providing the corresponding enantiomerically enriched sulfoximines. In contrast, stereoselective imidations of sulfoxides are rare. Finding an asymmetric imidation of racemic sulfoxides with concomitant in situ racemisation of the remaining sulfoxide enantiomer is a valuable target in this context. Despite all of the synthetic advances highlighted in this tutorial review, many improvements are still possible. Challenges to address relate to specific substitution pattern of the starting materials and products, the reaction conditions, which often are inapplicable for large-scale syntheses, and safety concerns, in particular, with respect to the imidating agents.

REFERENCES

Akai, S., & Kita, Y. (2007). Recent advances in Pummerer reactions. *Topics in Current Chemistry, 274*, 35–76.DOI: 10.1007/128_073

Anderson, K. K. (1979). Sulphonic Acids and their derivatives. In D. H. R. Barton, W. D. Ollis, & D.N. Jones (Eds.), *Comprehensive organic chemistry: the synthesis and reactions of organic compounds* (Vol 3: Sulphur, Selenium Silicon, Boron, Organomettalic Compounds) (pp. 331-362). Pergamon Press.

O'Connell, J. F., & Rapoport, H. (1992). 1- Benzenesulfonyl-and 1-p-toluenesulfonyl-3- methylimidazolium triflates: Efficient reagents for the preparation of arylsulfonamides and arylsulfonates. *The Journal of Organic Chemistry*, *57*(17). Advance online publication. DOI: 10.1021/jo00043a046

Banthorpe, D. V., Hughes, E. D., Ingold, C., Banthorpe, D. V., Bramley, R., & Thomas, J. A. (1964). Mechanism of benzidine and semidine rearrangements. Part XV. A Collective Discussion. *Journal of the Chemical Society*, 2864–2901. DOI: 10.1039/jr9640002864

Budavari, S. (Ed.). (1989). *The Merck index: An encyclopedia of chemicals, drugs, and biologicals* (11th ed.). Merck & Co., Inc.

Bur, S. K., & Padwa, A. (2004). The Pummerer reaction: Methodology and strategy for the synthesis of heterocyclic compounds. *Chemical Reviews*, *104*(5), 2401–2432. DOI: 10.1021/cr020090l PMID: 15137795

Bushby, S. R. M., & Hitchings, G. H. (1968). Trimethoprim, a sulphonamide potentiator. *British Journal of Pharmacology and Chemotherapy*, *33*(1), 72–90. DOI: 10.1111/j.1476-5381.1968.tb00475.x PMID: 5301731

Cairns, D. (2008). *Essentials of pharmaceutical chemistry* (3rd ed.). Pharmaceutical Press.

Carreño, M. C., Ribagorda, M., & Posner, G. H. (2002). Titanium-promoted stereoselective synthesis of hydroindolones from *p*-quinamines by domino conjugate additions. *Angewandte Chemie International Edition*, *41*(15), 2753–2755. DOI: 10.1002/1521-3773(20020802)41:15<2753::AID-ANIE2753>3.0.CO;2-M PMID: 12203474

Colobert, F., Tito, A., Khiar, N., Denni, D., Medina, M. A., Martin-Lomas, M., Ruano, J.-L. G., & Solladié, G. (1998). Enantioselective approach to polyhydroxylated compounds using chiral sulfoxides: Synthesis of enantiomerically pure *myo*-inositol and pyrrolidine derivatives. *The Journal of Organic Chemistry*, *63*(24), 8918–8921. DOI: 10.1021/jo9811569

Cox, J. R.Jr, & Dunn, M. F. (1963). The chemistry of O,N-diarylhydroxylamines – I. *Tetrahedron Letters*, *4*(15), 985–989. DOI: 10.1016/S0040-4039(01)90757-9

De, C. K., Pesciaioli, F., & List, B. (2013). Catalytic asymmetric benzidine rearrangement. *Angewandte Chemie International Edition*, *52*(35), 9293–9295. DOI: 10.1002/anie.201304039 PMID: 23832848

El-Sayed, N. S., El-Bendary, E. R., El-Ashry, S. M., & El-Kerdawy, M. M. (2011). Synthesis and antitumor activity of new sulfonamide derivatives of thiadiazolo[3,2-a]pyrimidines. *European Journal of Medicinal Chemistry*, *46*(9), 3714–3720. DOI: 10.1016/j.ejmech.2011.05.037 PMID: 21705114

Elgemeie, G. H., Azzam, R. A., & Elsayed, R. E. (2019). Sulfa drug analogs: New classes of N-sulfonyl aminated azines and their biological and preclinical importance in medicinal chemistry (2000–2018). *Medicinal Chemistry Research*, *28*(8), 1099–1131. DOI: 10.1007/s00044-019-02378-6

Feldman, K. S. (2006). Modern Pummerer-type reactions. *Tetrahedron*, *62*(21), 5003–5034. DOI: 10.1016/j.tet.2006.03.004

Fernandez, I., & Khiar, N. (2003). Recent developments in the synthesis and utilization of chiral sulfoxides. *Chemical Reviews*, *103*(9), 3651–3706. DOI: 10.1021/cr990372u PMID: 12964880

Gao, H., Ess, D. H., Yousufuddin, M., & Kürti, L. (2013). Transition-metal-free direct arylation: Synthesis of halogenated 2-amino-2B-hydroxy-1,1Bbiaryls and mechanism by DFT calculations. *Journal of the American Chemical Society, 135*(19), 7086–7089. DOI: 10.1021/ja400897u PMID: 23470200

García-Galán, M. J., Díaz-Cruz, M. S., & Barceló, D. (2008). Identification and determination of metabolites and degradation products of sulfonamide antibiotics. *Trends in Analytical Chemistry, 27*(11), 1008–1022. DOI: 10.1016/j.trac.2008.10.001

Gleckman, R., Alvarez, S., & Joubert, D. W. (1979). Drug therapy reviews: Trimethoprim-sulfamethoxazole. *American Journal of Hospital Pharmacy, 36*(7), 893–906. PMID: 382841

Guo, L., Liu, F., Wang, L., Yuan, H., Feng, L., Kürti, L., & Gao, H. (2019). Cascade approach to highly functionalized biaryls by a nucleophilic aromatic substitution with arylhydroxylamines. *Organic Letters, 21*(8), 2894–2898. DOI: 10.1021/acs.orglett.9b00927 PMID: 30942599

Harter, W. G., Albrect, H., Brady, K., Caprathe, B., Dunbar, J., Gilmore, J., Hays, S., Kostlan, C. R., Lunney, B., & Walker, N. (2004). The design and synthesis of sulfonamides as caspase-1 inhibitors. *Bioorganic & Medicinal Chemistry Letters, 14*(3), 809–812. DOI: 10.1016/j.bmcl.2003.10.065 PMID: 14741295

. Khiar, N., Fernández, I., & Alcudia, F. (1993). C2-symmetric bis-sulfoxides as chiral ligands in metal catalysed asymmetric Diels–Alder reactions. *Tetrahedron Letters, 34*(1), 123–126. (00DOI: 10.1016/S0040-4039

Kunieda, N., Nokami, J., & Kinoshita, M. (1976). β-Disulfoxides. II. The Preparation of Some Optically Active β-Disulfoxides. *Bulletin of the Chemical Society of Japan, 49*(1), 256–259. DOI: 10.1246/bcsj.49.256

Li, G.-Q., Gao, H., Keene, C., Devonas, M., Ess, D. H., & Kürti, L. (2013). Organocatalytic aryl–aryl bond formation: An atroposelective [3,3]-rearrangement approach to BINAM derivatives. *Journal of the American Chemical Society, 135*(20), 7414–7417. DOI: 10.1021/ja401709k PMID: 23659490

Özbek, N., Katırcıoğlu, H., Karacan, N., & Baykal, T. (2007). Synthesis, characterization and antimicrobial activity of new aliphatic sulfonamide. *Bioorganic & Medicinal Chemistry*, *15*(15), 5105–5109. DOI: 10.1016/j.bmc.2007.05.037 PMID: 17544281

Patai, S., Rappoport, H., & Sterling, J. J. (1988). *The chemistry of sulfones and sulphoxides*. Wiley.

Scozzafava, A., Owa, T., Mastrolorenzo, A., & Supuran, C. T. (2003). Anticancer and antiviral sulfonamides. *Current Medicinal Chemistry*, *10*(11), 925–953. DOI: 10.2174/0929867033457647 PMID: 12678681

Sheradsky, T., & Salemnick, G. (1971). The benzidine-like rearrangement of N,O-diarylhydroxylamines. *Tetrahedron Letters*, *12*(7), 645–648. DOI: 10.1016/S0040-4039(01)96519-0

Stranix, B. R., Lavallée, J.-F., Sévigny, G., Yelle, J., Perron, V., LeBerre, N., Herbart, D., & Wu, J. J. (2006). Lysine sulfonamides as novel HIV-protease inhibitors: Nε-acyl aromatic α-amino acids. *Bioorganic & Medicinal Chemistry Letters*, *16*(13), 3459–3462. DOI: 10.1016/j.bmcl.2006.04.011 PMID: 16644213

Thomas, G. (2007). *Medicinal chemistry* (2nd ed.). John Wiley & Sons.

Yanagi, T., & Yorimitsu, H. (2021). Recent development of biaryl synthesis through sigmatropic rearrangements. *Journal of Synthetic Organic Chemistry, Japan*, *79*(5), 427–438. DOI: 10.5059/yukigoseikyokaishi.79.427

Yuan, H., Du, Y., Liu, F., Guo, L., Sun, Q., Feng, L., & Gao, H. (2020). Tandem approach to NOBIN analogues from arylhydroxylamines and diaryliodonium salts via [3,3]-sigmatropic rearrangement. *Chemical Communications (Cambridge)*, *56*(59), 8226–8229. DOI: 10.1039/D0CC02919J PMID: 32555844

Zhang, J.-W., Qi, L.-W., Li, S., Xiang, S.-H., & Tan, B. (2020). Direct construction of NOBINs via domino arylation and sigmatropic rearrangement reactions. *Chinese Journal of Chemistry*, *38*(12), 1503–1514. DOI: 10.1002/cjoc.202000358

Chapter 7
Isomerism in Organic Compounds:
Diversity in Molecular Structure and Function

Rabab M. Nasser
https://orcid.org/0000-0003-2806-1051
Egyptian Petroleum Research Institute, Egypt

ABSTRACT

Isomerism is a fundamental concept in organic chemistry that describes the existence of compounds with the same molecular formula but different structural or spatial arrangements of atoms. This chapter explores the two primary categories of isomerism: constitutional (structural) isomerism and stereoisomerism. Within constitutional isomerism, various types such as chain, positional, functional, metameric, and tautomeric isomerism are discussed, highlighting their distinct characteristics and examples. Stereoisomerism is further divided into geometric and optical isomerism, emphasizing the significance of chirality and molecular symmetry. The implications of isomerism extend beyond theoretical chemistry, impacting fields like drug design and molecular synthesis. Understanding isomerism not only enhances our grasp of molecular diversity but also informs practical applications in various scientific disciplines.

DOI: 10.4018/979-8-3693-6473-4.ch007

INTRODUCTION

Isomerism is a fundamental concept in organic chemistry that refers to the phenomenon where two or more compounds share the same molecular formula but differ in their structural arrangement or spatial orientation of atoms. These compounds, known as isomers, can exhibit distinct physical and chemical properties despite having identical molecular compositions. This diversity in structure leads to a variety of behaviors and functionalities in chemical reactions and interactions, making isomerism a critical area of study in the field of organic chemistry, (Roos & Roos, 2015).

The importance of isomerism in organic chemistry cannot be overstated. Isomers can profoundly influence the characteristics of a compound, including its reactivity, boiling and melting points, solubility, and biological activity. For instance, in pharmaceuticals, the efficacy and safety of a drug can be significantly affected by its isomeric form. One isomer may exhibit therapeutic effects, while another could be inactive or even harmful. Understanding isomerism is therefore essential for chemists when designing and synthesizing new compounds, particularly in drug development and materials science, (Wermuth, 2015).

Moreover, the study of isomerism enhances our comprehension of molecular diversity and complexity, providing insights into how variations at the molecular level can lead to a wide range of chemical behaviors. Through the exploration of different types of isomers—structural isomers and stereoisomers—scientists can unlock new possibilities for innovation and application in various scientific fields.

TYPES OF ISOMERISM

Constitutional (Structural) Isomerism

Constitutional (structural) isomerism arises when atoms within a molecule are arranged in different orders. They are organic compounds that have the same molecular formula but different structural formula. Structural isomers usually show different physical and chemical properties.

Constitutional isomerism is vital in the study of organic compounds because it provides insights into how structural variations impact chemical properties and reactivity. By understanding the different types of constitutional isomers, chemists can predict how compounds will behave in reactions, their physical properties, and their potential applications in various fields, including pharmaceuticals, materials science, and chemical engineering, (Datta & Limpanuparb, 2021; Habler et al., 2024; Proniewicz & Lombardi, 2021).

Types of Structural Isomerism

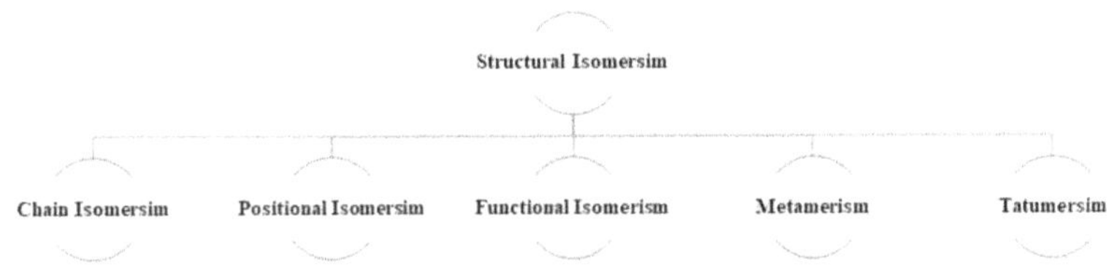

Chain Isomerism. The chain isomerism arises due to *different arrangements of carbon atoms leading to linear and branched chains*. The chain isomers are organic compounds that *have the same molecular formula but different types of chains* i.e., linear and branched. The chain isomers have almost similar chemical properties but different physical properties (Ye et al., 2023). For example, the branched chain isomers have lower boiling points than that of their linear counterparts. It is because the linear ones have more surface area of contact and *hence the intermolecular forces of attraction are maximum than the branched one*.

Butane and isobutane are chain isomers with molecular formula, C_4H_{12}. n-butane, is a linear chain isomer, while isobutane is a branched isomer.

butane isobutane

Positional Isomerism. The positional isomerism arises due to different positions of side chains, substituents, functional groups, double bonds, triple bonds etc., on the parent chain. Positional isomers are organic compounds that have the same molecular formula but differ in the position of functional group, (Che et al., 2024; Dey et al., 2021; Taiwo et al., 2024).

1) Propyl chloride and isopropyl chloride are the positional isomers with the molecular formula, C_3H_7Cl. These isomers arise due to difference in the **position of the chloro group** on the main chain.

1-chloropropane 2-chloropropane

2) But-1-ene and but-2-ene the **position of double bond** is different but they have same molecular formula, C_4H_8.

(*E*)-but-2-ene but-1-ene

3) The **functional group**, -OH is at different positions on the main chain of the following alcohols. They have the same molecular formula, C_3H_7OH.

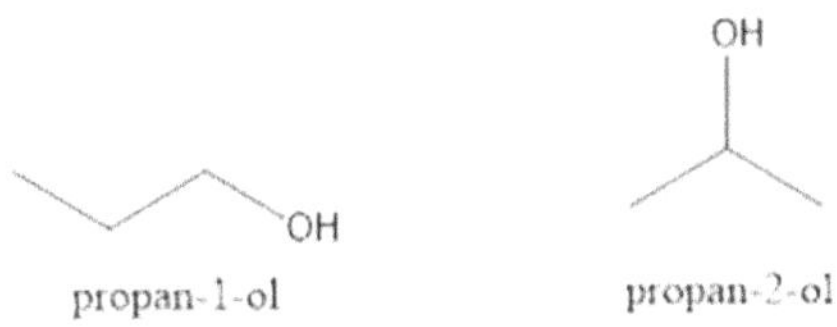

4) ortho xylene, meta xylene and para xylene, the relative positions of **methyl groups on the benzene ring** are different so they are positional isomers.

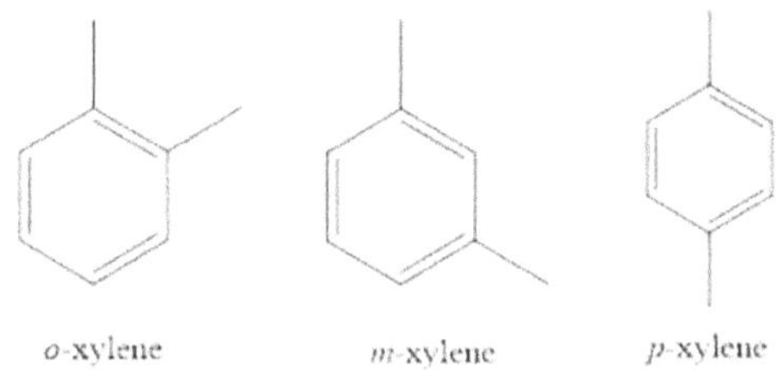

Functional Isomerism. The functional isomerism arises due to the presence of different functional groups. The functional isomers are organic compounds that have the same molecular formula but differ in the type of functional group, i.e. possess different functional groups. The functional isomers show different physical as well as chemical properties, (Ye et al., 2023).

1) The **dimethyl ether** is the functional isomer of **ethanol**. Both have the same molecular formula, C_2H_6O. However, they have different functional groups.

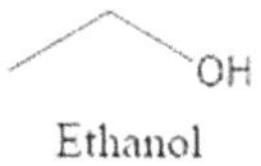

Ethanol

Di methyl ether

2) Propanal is a **functional isomer** of acetone. They have the same molecular formula, C_3H_6O but different functional groups. Propanal contains aldehyde, -CHO functional group, whereas acetone has ketone, -C=O functional group.

Propanal

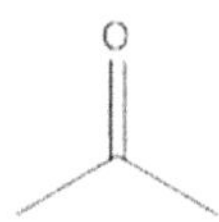

Acetone

3) Both Acetic acid and methyl formate are also **functional isomers**. They have the same molecular formula, $C_2H_4O_2$. Acetic acid is a carboxylic acid with -COOH group whereas methyl formate is an ester with $-COOCH_3$ group.

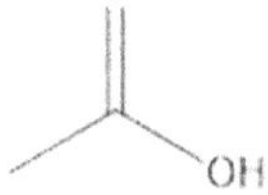

Acetic acid

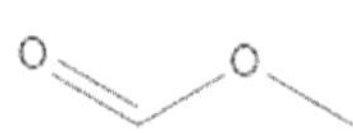

Methyl formate

Metamerism. Metamerism arises when different alkyl groups are attached to the same functional group.

1) The following metamers contain the **ether functional group**. However, they *differ by the nature of alkyl groups attached to the oxygen atom* (Ye *et al.*, 2023).

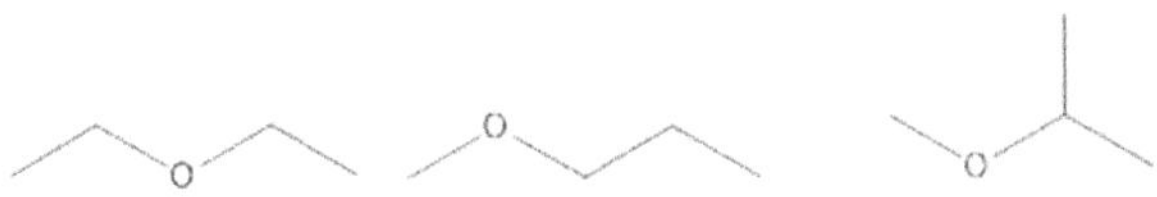

Diethyl ether | Methyl Propyl ether | Methyl Isopropyl ether

2) Metamerism is also possible in **amines** as shown below.

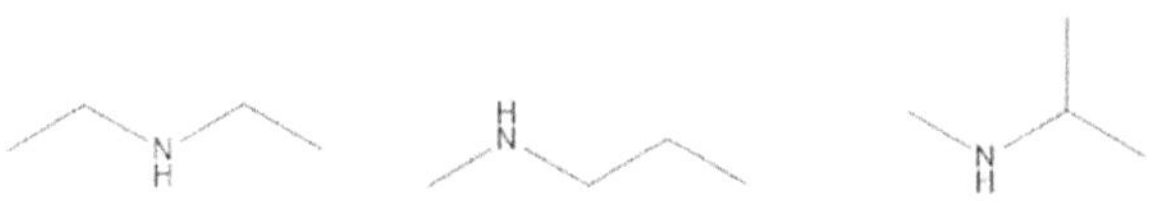

Diethyl amine | methyl propyl amine | Methyl Isopropyl amine

Tautomerism. Tautomerism (It is also called desmotropism or kryptotropism or prototropy or allelotropism.), refers to the dynamic equilibrium between two compounds with same molecular formula. It is most often a special case of functional group isomerism. In general, the tautomers have different functional groups and exist in dynamic equilibrium with each other due to a rapid inter-conversion from one form to another, (Ye *et al.*, 2023).

It is very important to **note that the tautomers are not the resonance structures of the same compound**.

1) **Keto-enol tautomerism**: The carbonyl compounds containing at least one α-hydrogen atom exhibit keto-enol tautomerism. The carbonyl group can be converted to enol form due to transfer of one of the α-hydrogen onto the oxygen atom.

Note: An alcoholic group on C=C is called enol. It is an alkene alcohol.

The % composition of keto enol tautomeric mixture depends on the relative stabilities of these two forms. In general, the keto form is the low energy form and is more stable than the enol form. However, the enol form is also stable in certain cases due to other stability factors.

The dynamic equilibrium between a keto form and an enol form exhibited by acetone is shown below. The tautomers are formed due to complete transfer of hydrogen. However, the relatively stable keto form exists in higher percentage (about 99%).

2) **The nitro-aci tautomerism**: It is exhibited by nitro compounds containing at least one α-hydrogen. In this case also, the hydrogen atom is transferred completely from one atom to another during the conversion of nitro form to aci form and vice versa, e.g. the nitro-aci tautomerism exhibited by nitromethane is shown below.

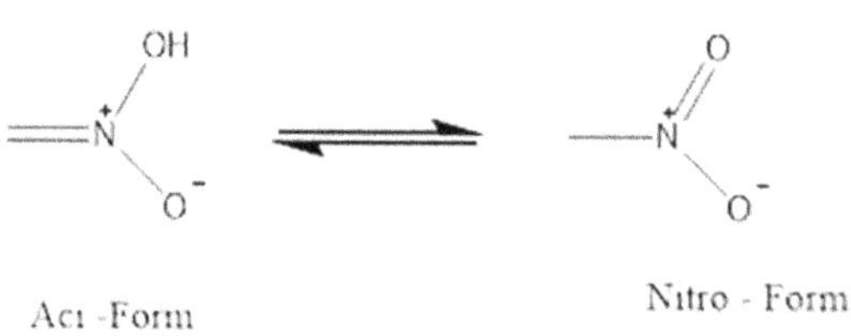

STEREOISOMERISM IN ORGANIC COMPOUNDS

Stereoisomerism is a form of isomerism that occurs when compounds have the same molecular formula and the same connectivity of atoms but differ in the spatial arrangement of those atoms or groups in three-dimensional space. This type of isomerism is essential in organic chemistry because even small changes in spatial arrangement can lead to significant differences in the properties and reactivity of the compounds. Stereoisomers can exhibit distinct physical and chemical behaviors, making their study crucial in various applications, including pharmaceuticals and materials science (Ye et al., 2023). ***Stereoisomers are organic compounds that have the same molecular formula but differ in the arrangement of atoms or groups in space***. It is broadly divided into:

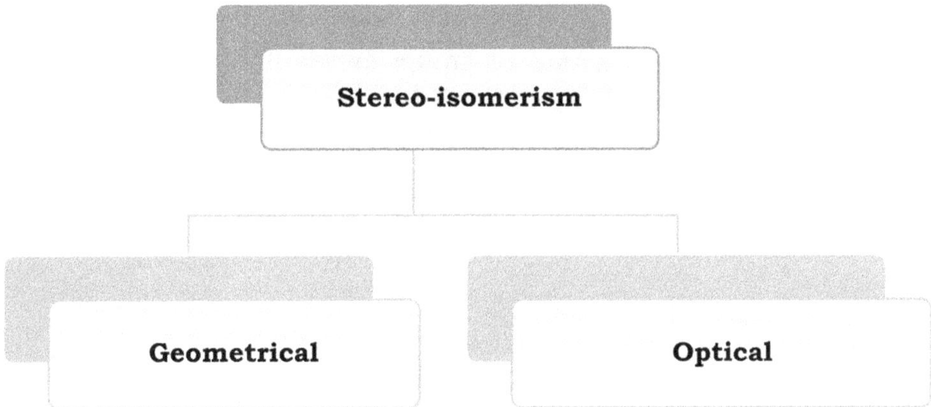

The stereo isomers have the same structural formula i.e., same connectivity order between atoms. However, the atoms or groups are arranged differently in space.

Geometrical Isomerism

Geometrical isomerism arises when atoms or groups are arranged differently in space due to restricted rotation of a bond or bonds in a molecule. Geometric isomers are organic compounds that have the same molecular formula but differ in the arrangement of atoms or groups in space, (Ye *et al.*, 2023).

Geometrical isomerism requires:

1. A carbon-carbon double bond C=C.
2. Each carbon atom of the double bond carries two different atoms or groups.

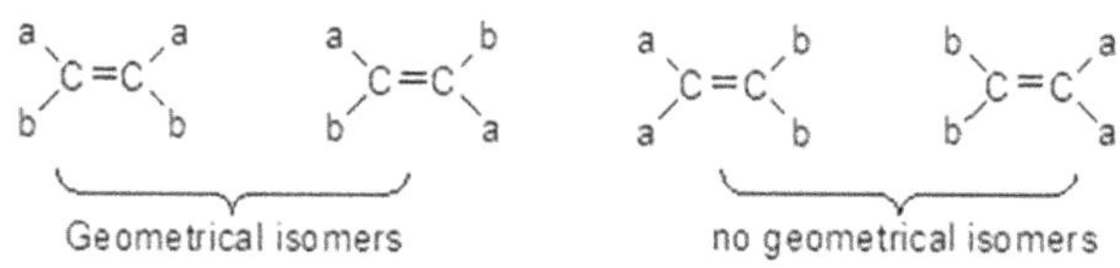

When the two similar atoms or groups are on the **same side** of the double bond the isomer is said to be ***cis* isomer**.

When the two similar atoms or groups are on the **opposite sides** of the double bond the isomer is said to be ***Trans-* isomer**.

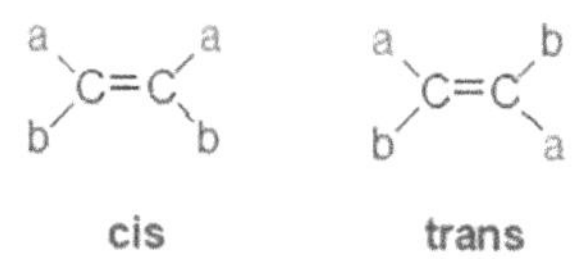

Two different spatial arrangements of methyl groups about a double bond in 2-butene give rise to the following geometrical isomers.

i.e., cis-2-butene and trans-2-butene

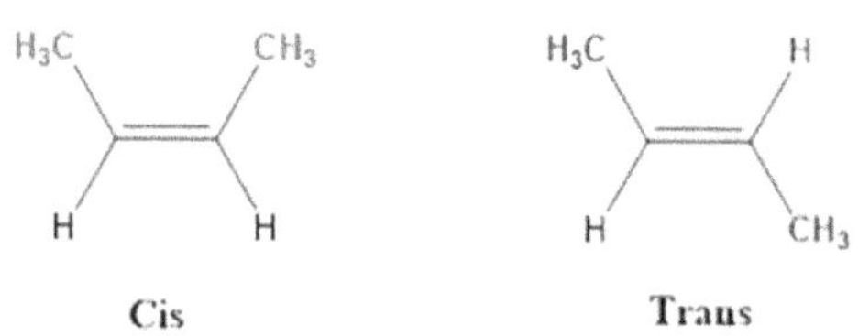

In the ***cis* isomer**, the two methyl groups are arranged on the same side of a double bond. Whereas in the ***Trans* isomer**, they are on the opposite side.

Table 1.

Consider the following examples The first double bond is Trans because the two hydrogen atoms are on opposite sides of the double bond. The second double bond is Cis because the two hydrogen atoms are on the same side of the double bond.	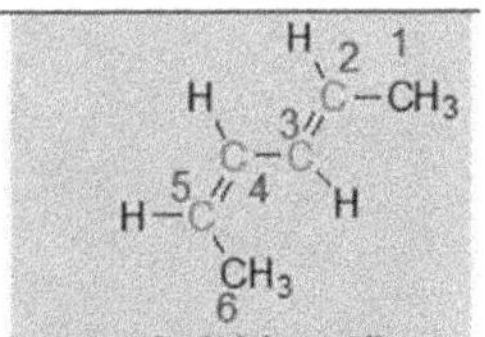 ***trans,cis*-2,4-hexadiene**
The first double bond is Cis because the two hydrogen atoms are on the same side of the double bond. The second double bond is Trans because the two hydrogen atoms are on opposite sides of the double bond.	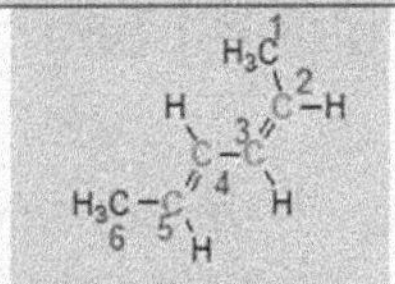 *cis, trans*-2,4-hexadiene
The first double bond is Cis because the two hydrogen atoms are on same side of the double bond. The second double bond is Cis because the two hydrogen atoms are on the same side of the double bond.	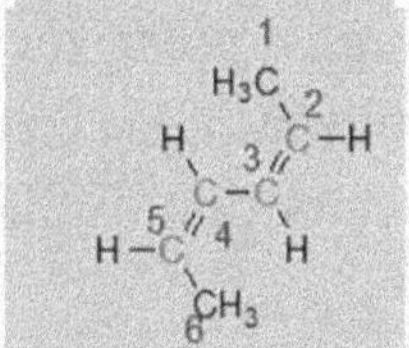 ***cis, cis*-2,4-hexadiene**
The first double bond is Trans because the two hydrogen atoms are on opposite sides of the double bond. The second double bond is Trans because the two hydrogen atoms are on opposite sides of the double bond.	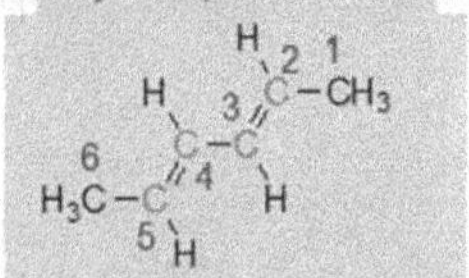 ***trans, trans*-2,4-hexadiene**

Optical Isomerism

The optical isomerism arises when atoms or groups are arranged differently in space and have different effects on the plane polarized light.

Optical isomers are organic compounds that have the same molecular formula but differ in the arrangement of atoms or groups in space and rotate the plane polarized light to right or left.

The organic compound which rotates the plane polarized light is called an optically active compound or chiral compound.

The optically active compound is an organic compound which contains at least one chiral carbon atom or stereogenic carbon or stereocenter.

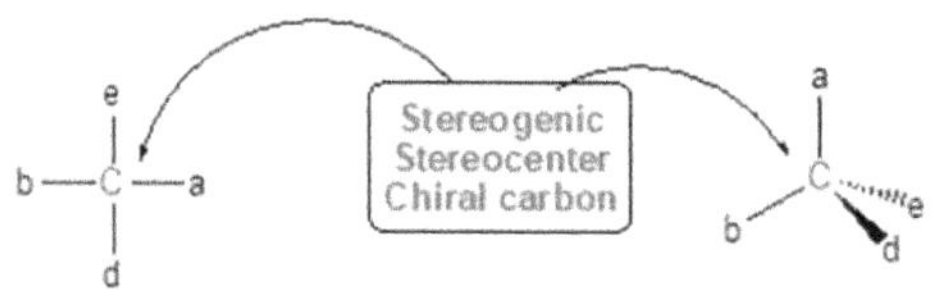

Optically active molecule

What is plane-polarized light?

Ordinary light consists of electromagnetic waves that oscillate in all planes perpendicular to the direction in which the light travels. Passing light through a polarizer allows light in only one plane to come through. This is **a plane-polarized light** (or simply **polarized light**), and it has an electric vector that oscillates in a single plane.

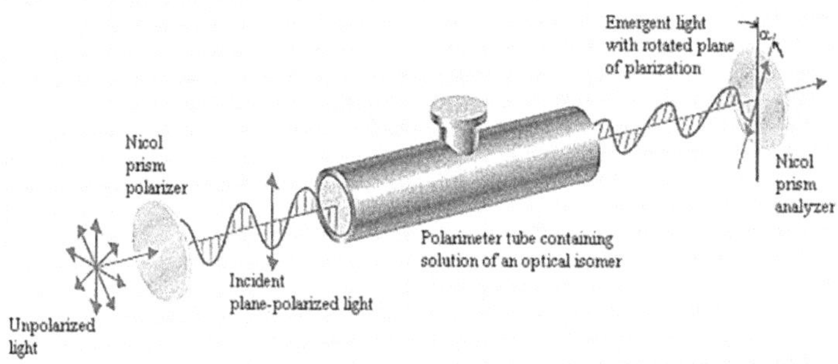

Polarimeter. A **polarimeter** is an instrument that allows plane-polarized light to travel through a sample tube containing an organic compound. After the light exits the sample tube, an analyzer slit is rotated to determine the direction of the plane of the polarized light exiting the sample tube. There are two possible results. With **achiral compounds,** the light exits the sample tube *unchanged,* and the plane of the polarized light is in the same position it was before entering the sample tube.

A compound that does not change the plane of polarized light is said to be ***optically inactive.***

The compound that rotates the plane-polarized light in the positive direction, or clockwise, is called dextrorotary [(+), or **d**]. While the compound that rotates the plane-polarized light in the negative direction, or counterclockwise, is called levorotary [(-), or **l**].

When both d and l isomers are present in equal amounts, the mixture is called a racemic mixture.

d*,** and ***l isomers of an organic are called **enantiomers** because they rotate the plane-polarized light in **opposite directions**, one to the **right** and the other to the **left**.

Enantiomers are **not superimposable** and they are mirror image to each other.

Figure 20. Chiral and achiral objects

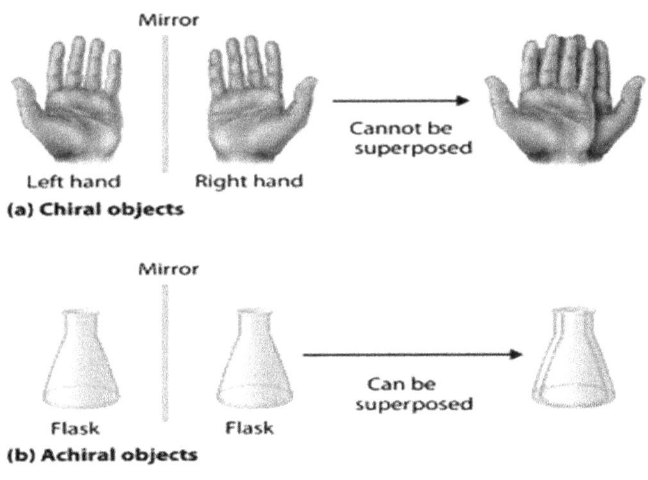

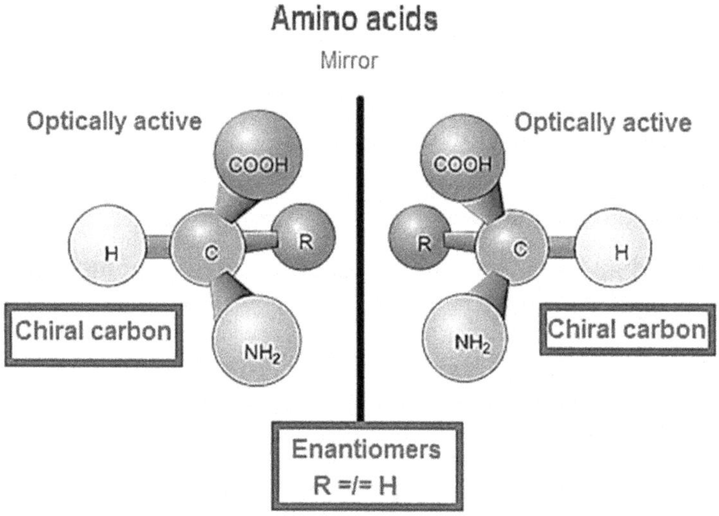

Elements of Symmetry

Symmetry is a fundamental concept in chemistry that describes the balanced and harmonious arrangement of atoms within a molecule. Understanding the elements of symmetry is crucial for analyzing molecular structures, predicting chemical behavior, and determining optical activity. In organic chemistry, four primary elements of symmetry are commonly referenced: the proper axis of symmetry, plane of symmetry, center of symmetry, and alternating axis of symmetry. Each element plays a distinct role in characterizing the symmetry of a molecule.

Proper Axis of Symmetry (Cn)

A proper axis of symmetry, denoted as Cn, is an imaginary line around which a molecule can be rotated by a specific angle (360°/n) to produce an indistinguishable arrangement.

- **C1**: No symmetry; the molecule is not superimposable on itself upon any rotation.

- **C2**: A molecule can be rotated 180° around the axis, yielding a configuration that is indistinguishable from the original.
- **C3, C4, C5, etc.**: These denote higher-order axes where n represents the number of identical arrangements obtained through rotation. For example, a C3 axis allows for 120° rotations.

Examples of molecules with proper axes include methane (C4) and benzene (C6), which exhibit significant symmetry in their structures.

Plane of Symmetry (σ)

A plane of symmetry is an imaginary plane that divides a molecule into two mirror-image halves. If a molecule can be reflected across this plane and produce an identical configuration, it possesses a plane of symmetry.

- **σv (Vertical Plane)**: A plane that includes the principal axis of symmetry and bisects the molecule.
- **σh (Horizontal Plane)**: A plane that is perpendicular to the principal axis of symmetry.
- **σd (Dihedral Plane)**: A plane that bisects the angles between two Cn axes.

Molecules with planes of symmetry exhibit a form of balance that can influence their physical and chemical properties. For instance, the presence of a plane of symmetry can render a molecule achiral.

Center of Symmetry (i)

A center of symmetry, also known as an inversion center, is a point within a molecule where any line drawn from this point to an atom will find an equivalent atom at an equal distance on the opposite side.

- If a molecule has a center of symmetry, it is symmetrical in all directions from this central point. This means that for every atom at position (x, y, z), there is an equivalent atom at (-x, -y, -z).

Molecules such as carbon dioxide (CO_2) exhibit a center of symmetry, as the arrangement of atoms on either side of the central carbon atom is perfectly mirrored.

Alternating Axis of Symmetry (Sn)

An alternating axis of symmetry, denoted as Sn, combines elements of rotation and reflection. For a molecule to have an Sn axis, it must exhibit symmetry through a combination of a rotation followed by a reflection across a plane that is perpendicular to the axis of rotation.

- **S2**: This indicates a 180° rotation followed by reflection.
- **S3, S4, etc.**: These denote higher-order alternating axes.

The alternating axis of symmetry is particularly relevant in complex molecules where a simple rotation or reflection does not fully describe the symmetry of the structure.

Importance of Symmetry Elements. The elements of symmetry are essential for several reasons:

1. **Predicting Optical Activity**: Chirality and optical isomerism are closely linked to the presence or absence of symmetrical elements. Molecules lacking a plane of symmetry and a center of symmetry are typically chiral and can exhibit optical activity.
2. **Characterizing Molecular Structures**: Understanding the symmetry of a molecule allows chemists to predict its behavior in chemical reactions, including reactivity and selectivity.
3. **Facilitating Spectroscopic Analysis**: Symmetry considerations play a significant role in vibrational spectroscopy, helping to identify active and inactive molecular vibrations based on symmetrical rules.

The elements of symmetry provide a framework for understanding the arrangement of atoms within molecules. By analyzing these symmetry elements, chemists can gain insights into molecular behavior, reactivity, and the physical properties of organic compounds. This understanding is foundational for advancing research in fields such as organic synthesis, materials science, and drug development.

Chirality

The property of any molecule of being nonsuperimposable on its mirror image is called chirality.

A chiral center also called stereocenter or stereogenic center is an atom bonded tetrahedrally to four different atoms or groups (also called ligands). *A chiral center is usually a* ***Carbon*** *atom, but may be* ***N, P, S, and Si****...etc. Achiral center is usually indicated with an asterisk (*).* If the chiral center is carbon, it's called ***chiral carbon***.

Chiral carbon must be:

1. **Tetrahedral "sp3"**.
2. Attached to different four groups.

What's the most common feature which makes a compound chiral?

A molecule may be chiral not only due to the presence of a chiral center, but also due to the presence of a chiral axis, a chiral plane and helicity.

Organic compounds become optically active if they contain:

1. ***Chiral center***, or
2. ***Chiral axis***,
3. ***Doesn't contain element of symmetry***.

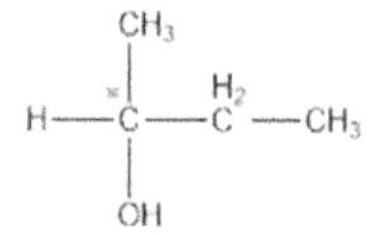

butan-2-ol

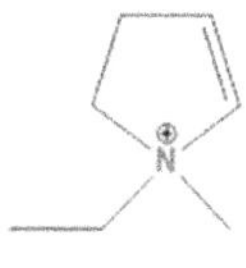

Quaternary nitrogen is a chiral center

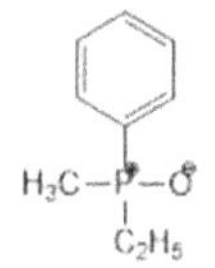

Phosphine oxide

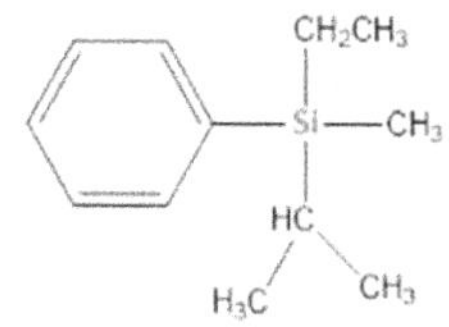

ethyl(isopropyl)(methyl)(phenyl)silane

Chiral compound Qualifications:

1. The compound and its image are not superimposable,
2. The compound and its image are called "enantiomers",
3. Has no element of symmetry?

What do you mean by a chiral and an achiral molecule?

If the **mirror image of a molecule is not superimposable** on the original, the molecule is called **a chiral molecule**. On the other hand, if the mirror image of a molecule is superimposable on the original, the molecule is called an **achiral molecule**. Ex: Lactic acid is a chiral molecule, whereas an ethanol molecule is an achiral molecule.

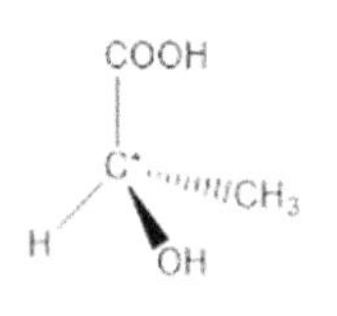

A stereoisomer of lactic acid
(a chiral molecule)

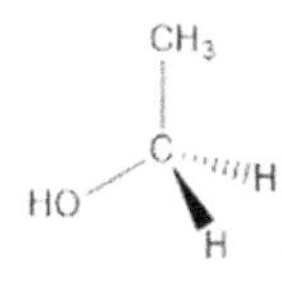

Ethanol
(an achiral molecule)

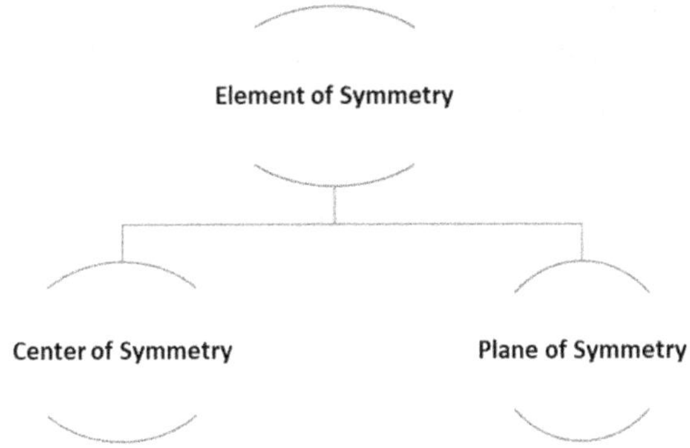

Plane of symmetry:

1. It's a mirror plane that ***cuts a molecule into two halves***.
2. ***One half is a reflection of the other half.***

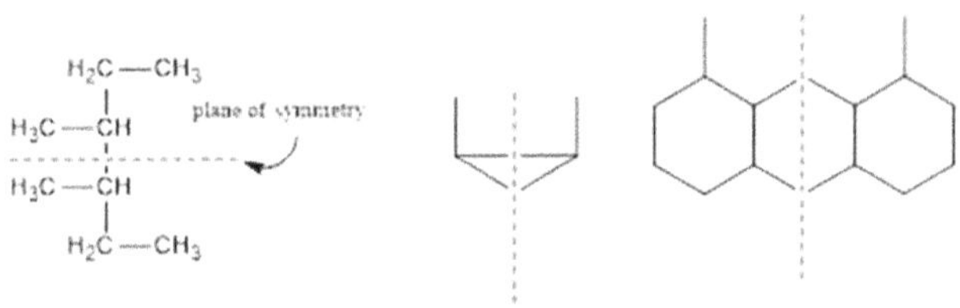

Center of Symmetry.

It's the ***point*** in the center of molecule to which a line can be drawn from any atom such that when extended an equal distance past the center, the line meets another atom of the same kind.

Figure 26. Center of symmetry

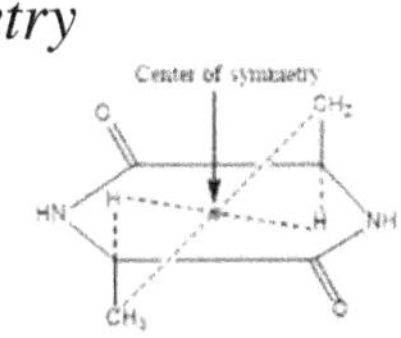

The compound that contains **a single tetrahedral carbon atom with four different substituents** (**Chiral carbon**) could exist in **two forms** that were **mirror images** of each other.

Ex., consider the CHFClBr molecule, for example, which contains four different substituents on a **tetrahedral carbon** atom.

The figure below shows one possible arrangement of these substituents and the mirror image of this structure. By convention,

1. **Solid lines** are used to represent bonds that lie in the **plane of the paper**.
2. **Wedges** are used for bonds that come **out of the plane** of the paper toward the viewer.
3. **Dashed lines** describe bonds that go behind the paper.

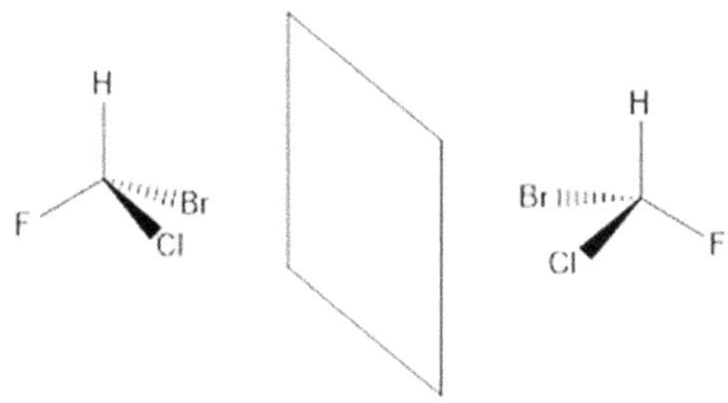

If we rotate the molecule on the right by 180° around the CH bond we get the structure shown on the right in the figure below.

These structures are different because they cannot be superimposed on each other, as shown in the figure below.

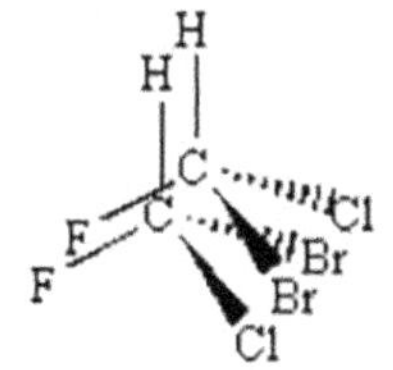

Problem (1):

Which of the following compounds would form enantiomers because the molecule is chiral?

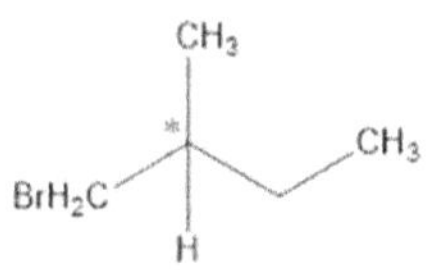

1-bromo-2-methylbutane

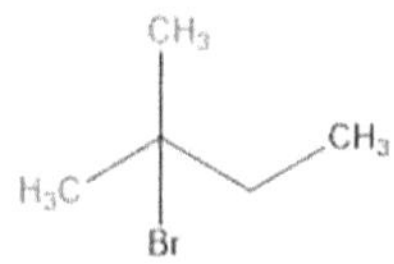

2-bromo-2-methylbutane

Answer:

The second carbon atom in 2-bromo-2-methylbutane contains two identical CH_3 substituents. As a result, this compound is ***achiral*** and does not form enantiomers.

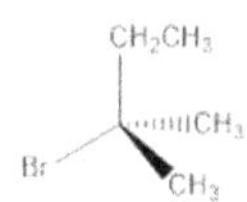

2 bromo 2 methylbutane

The second carbon atom in 1-bromo-2-methylbutane carries four different substituents: H, Br, CH_3, and CH_2CH_3. As a result, this molecule is ***chira*l** and it forms **enantiomers**.

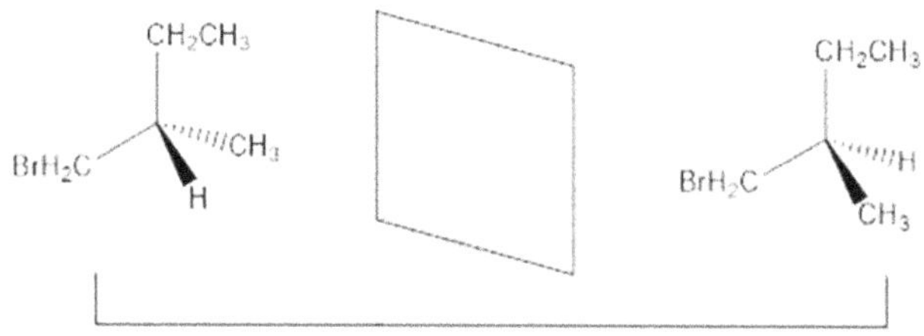

Enantiomers of 1-Bromo-2-methylbutane

Note: Every object with the possible exception of the vampires found on late-night TV has a mirror image. The relevant question is whether the mirror image can be superimposed on the source of the image. If it can, the object is not chiral (***achiral***). If it cannot, the object is chiral and it can exist as a **pair of stereoisomers**.

IMPLICATIONS OF ISOMERISM

Isomerism has profound implications in organic chemistry, significantly influencing both the physical and chemical properties of compounds. The arrangement of atoms and spatial orientation in isomers can lead to notable differences in boiling and melting points, solubility, and optical activity. For instance, linear isomers generally exhibit higher boiling points than their branched counterparts due to increased surface area and stronger intermolecular forces. Chemical properties also diverge among isomers, affecting their reactivity, selectivity in reactions, and overall stability; positional isomers, for example, can show varied reactivities based on the location of functional groups. These implications are particularly critical in drug design and synthesis, where the efficacy and safety of pharmaceuticals often depend on specific isomeric forms. Only one enantiomer of a chiral drug may be therapeutically active, underscoring the importance of careful design and synthesis of desired isomers to ensure effective treatment. Techniques such as asymmetric synthesis and chiral catalysis are employed to achieve high selectivity for specific isomers, thereby minimizing unwanted byproducts. Overall, understanding isomerism not only enriches our comprehension of organic compounds but also drives advancements in targeted drug action, synthetic methods, and innovative applications in biotechnology and materials science.

CONCLUSION

In conclusion, isomerism plays a crucial role in the field of organic chemistry, underpinning the diversity of molecular structures and their corresponding properties. The distinction between constitutional and stereoisomerism provides a comprehensive framework for understanding how variations in atom arrangement can lead to significant differences in behavior and function. As research continues to unveil new isomers and their applications, the relevance of isomerism in drug development, materials science, and biochemistry becomes increasingly evident. Future studies may focus on the exploration of novel isomers and their potential roles in innovative chemical processes, further expanding the boundaries of organic chemistry.

REFERENCES

Che, P., Chang, C., Buzzini, P., Stegemann, L., Kool, J., Davidson, J. T., & Kohler, I. (2024). Identification of synthetic cathinone positional isomers using electron activated dissociation mass spectrometry. *Analytica Chimica Acta*, *1319*(July), 342949. DOI: 10.1016/j.aca.2024.342949 PMID: 39122291

Datta, S., & Limpanuparb, T. (2021). Steric effectsvs.electron delocalization: A new look into the stability of diastereomers, conformers and constitutional isomers. *RSC Advances*, *11*(34), 20691–20700. DOI: 10.1039/D1RA02877D PMID: 35479364

Dey, A., Ramlal, V. R., Sankar, S. S., Kundu, S., Mandal, A. K., & Das, A. (2021). Self-assembled cationic organic nanosheets: Role of positional isomers in a guanidinium-core for efficient lithium-ion conduction. *Chemical Science (Cambridge)*, *12*(41), 13878–13887. DOI: 10.1039/D1SC04017K PMID: 34760173

Habler, K., Rexhaj, A., Adling-Ehrhardt, M., & Vogeser, M. (2024). Understanding isotopes, isomers, and isobars in mass spectrometry. *Journal of Mass Spectrometry and Advances in the Clinical Lab*, *33*(July), 49–54. DOI: 10.1016/j.jmsacl.2024.08.002 PMID: 39279892

Proniewicz, E., & Lombardi, J. R. (2021). Influence of excitation wavelength, solution pH, incubation time, absolute configuration (S and R) and constitutional isomerism (ortho-, meta- and para-) on the adsorption on the surface of Ag nanoparticles of N-benzylamino(boronphenyl) methylphosphonic ac. *Applied Surface Science*, *563*(May), 150341. DOI: 10.1016/j.apsusc.2021.150341

Roos, G., & Roos, C. (2015). Isomers and Stereochemistry. *Organic Chemistry Concepts*, 43–54. DOI: 10.1016/B978-0-12-801699-2.00003-1

Taiwo, B. J., Miller, A. H., Fielding, A. J., Sarker, S. D., van Heerden, F. R., & Fatokun, A. A. (2024). Ceibinin, a new positional isomer of mangiferin from the inflorescence of Ceiba pentandra (Bombacaceae), elicits similar antioxidant effect but no anti-inflammatory potential compared to mangiferin. *Heliyon, 10*(1), e23335. DOI: 10.1016/j.heliyon.2023.e23335 PMID: 38332887

Wermuth, C. G. (2015). Optical Isomerism in Drugs. In *The Practice of Medicinal Chemistry* (4th ed.). Elsevier Ltd., DOI: 10.1016/B978-0-12-417205-0.00018-3

Ye, F., Liao, C., Cheng, H., Zhang, J., Wang, H., Li, Y., Li, H., & Ming, J. (2023). Fundamentals of Organic Chemistry. In *Comprehensive Experiments for Materials Science and Engineering*. DOI: 10.1142/9789811274053_0003

Compilation of References

Abramovitch, R. (2011). *Reactive intermediates*. Springer.

Ahluwalia,V. K., &Parashar,R. K.(2011). Organic reaction mechanisms.

Akai, S., & Kita, Y. (2007). Recent advances in Pummerer reactions. *Topics in Current Chemistry*, *274*, 35–76.DOI: 10.1007/128_073

Anderson, K. K. (1979). Sulphonic Acids and their derivatives. In D. H. R. Barton, W. D. Ollis, & D.N. Jones (Eds.), *Comprehensive organic chemistry: the synthesis and reactions of organic compounds* (Vol 3: Sulphur, Selenium Silicon, Boron, Organomettalic Compounds) (pp. 331-362). Pergamon Press.

Anderson-Wile, A. M., Hall, K. E., & Celius, T. C. (2024). Scaffolded structure determination activities utilizing 2D NMR Spectroscopy of simple organic molecules. *Journal of Chemical Education*, *101*(3), 1356–1360. DOI: 10.1021/acs.jchemed.3c00882

Atherton, S. J., & Harriman, A. (1993). Time-resolved studies of intramolecular photoinduced electron transfer in donor-acceptor systems. *Journal of the American Chemical Society*, *115*(5), 1816–1822. DOI: 10.1021/ja00058a020

Atkins, P., & de Paula, J. (2014). *Atkins' Physical Chemistry* (10th ed.). Oxford University Press. DOI: 10.1093/acprof:oso/9780198769286.001.0001

Banthorpe, D. V., Hughes, E. D., Ingold, C., Banthorpe, D. V., Bramley, R., & Thomas, J. A. (1964). Mechanism of benzidine and semidine rearrangements. Part XV. A Collective Discussion. *Journal of the Chemical Society*, 2864–2901. DOI: 10.1039/jr9640002864

Benderskii, V. A., & Grebenshchikov, S. Y. (1994). Solvation dynamics and electron transfer in liquids. *Journal of Electroanalytical Chemistry (Lausanne, Switzerland)*, *375*(1–2), 29–44. DOI: 10.1016/0022-0728(94)03471-1

Bingham, N. M., Wright, J. S., Mathias, S. C., Douce, D., & Sears, P. (2024). Atmospheric solids analysis probe mass spectrometry: An easy bolt-on for the synthetic undergraduate teaching laboratory. *Rapid Communications in Mass Spectrometry*, *38*(5), e9689. DOI: 10.1002/rcm.9689 PMID: 38248910

Binstead, R. A., & Meyer, T. J. (1992). Thermodynamics and kinetics of transition-metal oxidation reactions in solution. *Journal of the American Chemical Society*, *114*(1), 173–186. DOI: 10.1021/ja00027a022

Blair, D. F., Ellis, W. R.Jr, Wang, H., Gray, H. B., & Chan, S. I. (1986). Electron transfer in cytochrome c oxidase. *The Journal of Biological Chemistry*, *261*(25), 11524–11537. DOI: 10.1016/S0021-9258(18)67749-1

Bögerhausen, O., Haumann, MJunge, W. (1996). Photosynthetic oxygen evolution: H/D isotope effects and the coupling between electron and proton transfer during transitions S2⟹3 and S3⟹S4→S0. *Berichte der Bunsengesellschaft fur Physikalische Chemie*, *100*(12), 1987–1992. 10.1002/bbpc.19961001210

Bromfield Lee, D., & Nelson, J. (2024). Modifying an organic chemistry esterification teaching lab to be accessible to blind and visually impaired (BVI) students. *Journal of Chemical Education*, *101*(5), 1891–1898. DOI: 10.1021/acs.jchemed.3c01081

Brudvig, G. W., Holden Thorp, H., & Crabtree, R. H. (1991). Probing the mechanism of water osidation in photosystem II. *Accounts of Chemical Research*, *24*(10), 311–316.

Bruice, P. Y. (2015). *Essential Organic Chemistry* (Global Edition).

Budavari, S. (Ed.). (1989). *The Merck index: An encyclopedia of chemicals, drugs, and biologicals* (11th ed.). Merck & Co., Inc.

Bur, S. K., & Padwa, A. (2004). The Pummerer reaction: Methodology and strategy for the synthesis of heterocyclic compounds. *Chemical Reviews, 104*(5), 2401–2432. DOI: 10.1021/cr020090l PMID: 15137795

Bushby, S. R. M., & Hitchings, G. H. (1968). Trimethoprim, a sulphonamide potentiator. *British Journal of Pharmacology and Chemotherapy*, *33*(1), 72–90. DOI: 10.1111/j.1476-5381.1968.tb00475.x PMID: 5301731

Caceres-Cortes, J., Falk, B., Mueller, L., & Dhar, T. G. M. (2024). Perspectives on nuclear magnetic resonance spectroscopy in drug discovery research. *Journal of Medicinal Chemistry*, *67*(3), 1701–1733. DOI: 10.1021/acs.jmedchem.3c02389 PMID: 38290426

Cairns, D. (2008). *Essentials of pharmaceutical chemistry* (3rd ed.). Pharmaceutical Press.

Carey, F. A., & Sundberg, R. J. (2007). *Advanced Organic Chemistry: Part A: Structure and Mechanisms*. Springer Science & Business Media. DOI: 10.1007/978-0-387-44899-2

Carreño, M. C., Ribagorda, M., & Posner, G. H. (2002). Titanium-promoted stereoselective synthesis of hydroindolones from *p*-quinamines by domino conjugate additions. *Angewandte Chemie International Edition, 41*(15), 2753–2755. DOI: 10.1002/1521-3773(20020802)41:15<2753::AID-ANIE2753>3.0.CO;2-M PMID: 12203474

Chang, R., & Goldsby, K. A. (2013). *Chemistry* (11th ed.). McGraw-Hill Education. DOI: 10.1036/0073402661

Che, P., Chang, C., Buzzini, P., Stegemann, L., Kool, J., Davidson, J. T., & Kohler, I. (2024). Identification of synthetic cathinone positional isomers using electron activated dissociation mass spectrometry. *Analytica Chimica Acta, 1319*(July), 342949. DOI: 10.1016/j.aca.2024.342949 PMID: 39122291

Clayden, J., Greeves, N., & Warren, S. (2012). *Organic Chemistry*. Oxford University Press. DOI: 10.1093/hesc/9780199270293.001.0001

Colobert, F., Tito, A., Khiar, N., Denni, D., Medina, M. A., Martin-Lomas, M., Ruano, J.-L. G., & Solladié, G. (1998). Enantioselective approach to polyhydroxylated compounds using chiral sulfoxides: Synthesis of enantiomerically pure *myo*-inositol and pyrrolidine derivatives. *The Journal of Organic Chemistry*, *63*(24), 8918–8921. DOI: 10.1021/jo9811569

Cox, J. R.Jr, & Dunn, M. F. (1963). The chemistry of O,N-diarylhydroxylamines – I. *Tetrahedron Letters*, *4*(15), 985–989. DOI: 10.1016/S0040-4039(01)90757-9

Datta, S., & Limpanuparb, T. (2021). Steric effectsvs.electron delocalization: A new look into the stability of diastereomers, conformers and constitutional isomers. *RSC Advances*, *11*(34), 20691–20700. DOI: 10.1039/D1RA02877D PMID: 35479364

De, C. K., Pesciaioli, F., & List, B. (2013). Catalytic asymmetric benzidine rearrangement. *Angewandte Chemie International Edition*, *52*(35), 9293–9295. DOI: 10.1002/anie.201304039 PMID: 23832848

Dey, A., Ramlal, V. R., Sankar, S. S., Kundu, S., Mandal, A. K., & Das, A. (2021). Self-assembled cationic organic nanosheets: Role of positional isomers in a guanidinium-core for efficient lithium-ion conduction. *Chemical Science (Cambridge)*, *12*(41), 13878–13887. DOI: 10.1039/D1SC04017K PMID: 34760173

Dorsheimer, J. R., & Rovis, T. (2023). Late-stage isotopic exchange of primary amines. *Journal of the American Chemical Society*, *145*(44), 24367–24374. DOI: 10.1021/jacs.3c09442 PMID: 37889497

Eberhart, A. J., & Procter, D. J. (2011). Nucleophilic ortho allylation of aryl and heteroaryl sulfoxides. *Organic Letters*, *13*(22), 5882–5885. DOI: 10.1021/ol202483e

Eberhart, A. J., & Procter, D. J. (2013). Nucleophilic ortho-allylation of pyrroles and pyrazoles: An accelerated Pummerer/thio-Claisen rearrangement sequence. *Organic Letters*, *15*(16), 3994–3997. DOI: 10.1021/ol4016734

Eberhart, A. J., & Procter, D. J. (2013). Nucleophilic ortho-propargylation of aryl sulfoxides: An interrupted Pummerer/allenyl thio-Claisen rearrangement sequence. *Angewandte Chemie International Edition*, *52*(15), 4008–4011. DOI: 10.1002/anie.201209544

Eberhart, A. J., & Procter, D. J. (2015). Sulfoxide-directed metal-free ortho-propargylation of aromatics and heteroaromatics. *Chemistry (Weinheim an der Bergstrasse, Germany)*, *21*(19), 7428–7434. DOI: 10.1002/chem.201500254

Elgemeie, G. H., Azzam, R. A., & Elsayed, R. E. (2019). Sulfa drug analogs: New classes of N-sulfonyl aminated azines and their biological and preclinical importance in medicinal chemistry (2000–2018). *Medicinal Chemistry Research*, *28*(8), 1099–1131. DOI: 10.1007/s00044-019-02378-6

El-Sayed, N. S., El-Bendary, E. R., El-Ashry, S. M., & El-Kerdawy, M. M. (2011). Synthesis and antitumor activity of new sulfonamide derivatives of thiadiazolo[3,2-a]pyrimidines. *European Journal of Medicinal Chemistry*, *46*(9), 3714–3720. DOI: 10.1016/j.ejmech.2011.05.037 PMID: 21705114

Felderhoff, M., Steller, I., Reyes-Arellano, A., Boese, R., & Sustmann, R. (1996). Synthesis and structures of diazomethane-derived cyclopropanes. *Advanced Materials*, *8*(5), 402–405. DOI: 10.1002/adma.19960080507

Feldman, K. S. (2006). Modern Pummerer-type reactions. *Tetrahedron*, *62*(21), 5003–5034. DOI: 10.1016/j.tet.2006.03.004

Fernandez, I., & Khiar, N. (2003). Recent developments in the synthesis and utilization of chiral sulfoxides. *Chemical Reviews*, *103*(9), 3651–3706. DOI: 10.1021/cr990372u PMID: 12964880

Gao, H., Ess, D. H., Yousufuddin, M., & Kürti, L. (2013). Transition-metal-free direct arylation: Synthesis of halogenated 2-amino-2B-hydroxy-1,1Bbiaryls and mechanism by DFT calculations. *Journal of the American Chemical Society*, *135*(19), 7086–7089. DOI: 10.1021/ja400897u PMID: 23470200

García-Galán, M. J., Díaz-Cruz, M. S., & Barceló, D. (2008). Identification and determination of metabolites and degradation products of sulfonamide antibiotics. *Trends in Analytical Chemistry*, *27*(11), 1008–1022. DOI: 10.1016/j.trac.2008.10.001

Giraudeau, P., & Remaud, G. (2020). Advanced NMR Methods for Position-Specific Isotope Analysis. In Yoshida, N., Gilbert, A., & Foriel, J. (Eds.), *Handbook of Isotopologue Biogeochemistry* (pp. 1–31). Springer Nature Singapore.DOI: 10.1007/978-981-10-7048-8_12-1

Glänzer, D., Pfeiffer, M., Ribar, A., Zeindl, R., Tollinger, M., Nidetzky, B., & Kreutz, C. (2024). Efficient synthetic access to stable isotope labeled pseudouridine phosphoramidites for RNA NMR Spectroscopy. *Chemistry (Weinheim an der Bergstrasse, Germany)*, e202401193. DOI: 10.1002/chem.202401193 PMID: 38652483

Gleckman, R., Alvarez, S., & Joubert, D. W. (1979). Drug therapy reviews: Trimethoprim-sulfamethoxazole.*American Journal of Hospital Pharmacy*, *36*(7), 893–906. PMID: 382841

Graige, M. S., Paddock, M. L., Bruce, J. M., Feher, G., & Okamura, M. Y. (1996). Mechanism of proton-coupled electron transfer for quinone (Q_B) reduction in reaction centers of *Rhodobacter sphaeroides.Journal of the American Chemical Society*, *118*(38), 9005–9016. PMC+1PMC+1 https://doi.org/10.1021/ja960056m

Guo, H. M., Wu, S., Niu, H. Y., Song, G., & Qu, G. R. (2013). Chemical Synthesis of Acyclic Nucleosides. In *Chemical Synthesis of Nucleoside Analogs* (pp. 103–162). Wiley., DOI: 10.1002/9781118498088.ch3

Guo, L., Liu, F., Wang, L., Yuan, H., Feng, L., Kürti, L., & Gao, H. (2019). Cascade approach to highly functionalized biaryls by a nucleophilic aromatic substitution with arylhydroxylamines. *Organic Letters*, *21*(8), 2894–2898. DOI: 10.1021/acs.orglett.9b00927 PMID: 30942599

Gust, D., Barrett, D., Harding, L. O., Moore, A. L., Moore, T. A., & Vermaas, W. F. J. (1988). Photochemical energy storage in reaction center model systems. *Journal of the American Chemical Society*, *110*(1), 321–323. DOI: 10.1021/ja00208a048

Gust, D., & Moore, A. L. (1993). Mimicking photosynthetic solar energy transduction. *Accounts of Chemical Research*, *26*(4), 198–205. DOI: 10.1021/ar00029a005

Gust, D., Moore, T. A., Moore, A. L., Gao, F., & Luttrull, D. K.. (1991). Photochemical charge separation in biomimetic molecular triads. *Journal of the American Chemical Society*, *113*(10), 3638–3649. DOI: 10.1021/ja00009a015

Habler, K., Rexhaj, A., Adling-Ehrhardt, M., & Vogeser, M. (2024). Understanding isotopes, isomers, and isobars in mass spectrometry. *Journal of Mass Spectrometry and Advances in the Clinical Lab*, *33*(July), 49–54. DOI: 10.1016/j.jmsacl.2024.08.002 PMID: 39279892

Harter, W. G., Albrect, H., Brady, K., Caprathe, B., Dunbar, J., Gilmore, J., Hays, S., Kostlan, C. R., Lunney, B., & Walker, N. (2004). The design and synthesis of sulfonamides as caspase-1 inhibitors. *Bioorganic & Medicinal Chemistry Letters*, *14*(3), 809–812. DOI: 10.1016/j.bmcl.2003.10.065 PMID: 14741295

Helmreich, R. F. (1953). *Mechanisms of Substitution and Elimination Reactions*. University of Colorado.

He, Z., & Procter, D. J. (2019). Pummerer chemistry of benzothiophene sulfoxides: Metal-free alkylation and arylation of benzothiophenes. *Phosphorus, Sulfur, and Silicon and the Related Elements*, *194*(7), 669–677. DOI: 10.1080/10426507.2019.1619603

Higuchi, K., & Tayu, M. (2021). The interrupted Pummerer reaction: Design of sulfoxides and their utility in organic synthesis. *Heterocycles*, *102*(3), 783–824. DOI: 10.3987/COM-20-14040

Hoenig, S. L. (2024). *Basic Chemical Concepts and Tables* (2nd ed.). CRC Press. DOI: 10.1201/9781003396512

Hoganson, C. W., & Babcock, G. T. (1997). A metalloradical mechanism for the generation of oxygen from water in photosynthesis. *Science*, *277*(5334), 1953–1956. PMC https://doi.org/10.1126/science.277.5334.1953

Hornback, J. M. (2005). *Organic chemistry*. Cengage Learning.

Huang, X., Klimczyk, S., & Maulide, N. (2012). Charge-accelerated sulfonium [3,3]-sigmatropic rearrangements. *Synthesis*, *44*(2), 175–183. DOI: 10.1055/s-0031-1289807

Huang, X., & Maulide, N. (2011). Sulfoxide-mediated α-arylation of carbonyl compounds. *Journal of the American Chemical Society*, *133*(21), 8510–8513. DOI: 10.1021/ja202591s

Huang, X., & Maulide, N. (2013). Sulfur(IV)-mediated transformations: From ylide transfer to metal-free arylation of carbonyl compounds. *Journal of the American Chemical Society*, *135*(20), 7312–7323. DOI: 10.1021/ja402710j

Hung, S.-C., Macpherson, A. N., Lin, S., Liddell, P. A., & Seely, G. R. (1995). Photocurrent generation by molecular triads in photoelectrochemical cells. *Journal of the American Chemical Society*, *117*(6), 1657–1658. DOI: 10.1021/ja00109a034

Ismail, F. M. D., Nahar, L., & Sarker, S. D. (2024). Prediction of structure based on spectral data using computational techniques. In Sarker, S. D., & Nahar, L. (Eds.), *Computational Phytochemistry* (2nd ed., pp. 255–300). Elsevier. DOI: 10.1016/B978-0-443-16102-5.00016-X

Jang, J., Park, J., & Shin, H. J. (2024). Terahertz spectral analysis: An in-depth exploration of spectroscopy approaches for ultrafast dynamics. *Current Applied Physics*, *59*, 197–207. DOI: 10.1016/j.cap.2023.11.015

Jirasek, M., Sharma, A., Bame, J. R., Mehr, S. H. M., Bell, N., Marshall, S. M., Mathis, C., MacLeod, A., Cooper, G. J. T., Swart, M., Mollfulleda, R., & Cronin, L. (2024). Investigating and quantifying molecular complexity using assembly theory and spectroscopy. *ACS Central Science*, *10*(5), 1054–1064. DOI: 10.1021/acscentsci.4c00120 PMID: 38799656

Kaiser, D., Oost, R., Neuhaus, J., & Maulide, N. (2019). Bond-forming and –breaking reactions at sulfur (IV): Sulfoxides, sulfonium salts, sulfur ylides, and sulfinate salts. *Chemical Reviews*, *119*(13), 8701–8780. DOI: 10.1021/acs.chemrev.9b00134

Kevill, D. N. (1959). *Substitution and Elimination Reactions in Non-hydroxylic Media.* University of London.

Kim, J., Hayashi, Y., Badr, S., Okamoto, K., Hakogi, T., Furukawa, H., Yoshikawa, S., Nakanishi, H., & Sugiyama, H. (2024). Kinetic study and model-based design space determination for a drug substance flow synthesis using an amination reaction via nucleophilic aromatic substitution. *Organic Process Research & Development*, *28*(5), 1793–1805. DOI: 10.1021/acs.oprd.3c00380

Kitagawa, T., & Ozawa, T. (1997). Resonance Raman spectroscopy of metalloporphyrins and heme proteins. In Lippard, S. J. (Ed.), *Progress in Inorganic Chemistry* (Vol. 45, pp. 431–479). Wiley., DOI: 10.1002/9780470166413.ch8

Klein, D. R. (2017). *Organic Chemistry.* John Wiley & Sons.http://books.google.ie/books?id=38aXDwAAQBAJ&pg=PA272&dq=substitution+and+elimination&hl=&cd=2&source=gbs_api

Kobatake, T., & Oshima, K. (2010). Reaction of 2-(2,2,2-trifluoroethylidene)-1,3-dithiane 1-oxide with ketones under Pummerer conditions and its application to the synthesis of trifluoromethyl-substituted five-membered heteroaromatics. *Angewandte Chemie International Edition*, *49*(13), 2340–2343. DOI: 10.1002/anie.200907016

Kobatake, T., & Oshima, K. (2010). Synthesis of 3-trifluoromethylbenzo[b]furans from phenols via direct ortho functionalization by extended Pummerer reaction. *Journal of the American Chemical Society*, *132*(33), 11838–11840. DOI: 10.1021/ja104317a

Koothradan, F. F., Jayarani, A., & Sivasankar, C. (2024). Synthesis of β-lactams through carbonylation of diazo compounds followed by the [2 + 2] cycloaddition reaction. *The Journal of Organic Chemistry*, *89*(7), 4294–4308. DOI: 10.1021/acs.joc.2c01484 PMID: 36469500

Kovalev, I. S., Zyryanov, G. V., Santra, S., Majee, A., Varaksin, M. V., & Charushin, V. N. (2022). Folic acid antimetabolites (antifolates): A Brief review on synthetic strategies and application opportunities. *Molecules (Basel, Switzerland)*, *27*(19), 6229. DOI: 10.3390/molecules27196229 PMID: 36234766

Krishtalik, L. I. (1970). Electrochemical mechanisms and kinetics of electron transfer reactions. In Delahay, P., & Tobias, C. W. (Eds.), *Advances in Electrochemistry and Electrochemical Engineering* (Vol. 7, pp. 283–339). Interscience.

Kunieda, N., Nokami, J., & Kinoshita, M. (1976). β-Disulfoxides. II. The Preparation of Some Optically Active β-Disulfoxides. *Bulletin of the Chemical Society of Japan*, *49*(1), 256–259. DOI: 10.1246/bcsj.49.256

Lavergne, J., & Junge, W. (1993). Proton release during the redox cycle of the water oxidase. *Photosynthesis Research*, *38*(3), 279–296. Groningen Research Portal+4SURFconext+4PMC+4 https://doi.org/10.1007/BF00046752

Levich, V. G. (1970). Physicochemical hydrodynamics. In H. Henderson & W. Yost (Eds.), *Physical Chemistry—An Advanced Treatise* (Vol. 9B, pp. 985–1074). Academic Press.

Liebman, J. F., & Greenberg, A. (1988). *Structure and Reactivity*. Wiley-VCH.

Li, G.-Q., Gao, H., Keene, C., Devonas, M., Ess, D. H., & Kürti, L. (2013). Organocatalytic aryl–aryl bond formation: An atroposelective [3,3]-rearrangement approach to BINAM derivatives. *Journal of the American Chemical Society*, *135*(20), 7414–7417. DOI: 10.1021/ja401709k PMID: 23659490

Lima, L. M., Silva, B. N. M., Barbosa, G., & Barreiro, E. J. (2020). β-lactam antibiotics: An overview from a medicinal chemistry perspective. *European Journal of Medicinal Chemistry*, *208*, 112829. DOI: 10.1016/j.ejmech.2020.112829 PMID: 33002736

Linne, M. A. (2024). *Spectroscopic Measurement: An Introduction to the Fundamentals*. Elsevier., DOI: 10.1016/C2022-0-00036-2

Ma, J., Ye, Q., Green, R. A., Gurak, J., Ayers, S., Huang, Y., & Miller, S. A. (2024). Overcoming NMR line broadening of nitrogen containing compounds: A simple solution. *Magnetic Resonance in Chemistry*, *62*(3), 198–207. DOI: 10.1002/mrc.5432 PMID: 38258438

McMurry, J. (2003). *Fundamentals of Organic Chemistry*. Brooks Cole.

Meyer, B., Schlodder, E., Dekker, J. P., & Witt, H. T. (1989). O_2 evolution and Chl a^{+}II ($P680^{+}$) reduction in photosystem II: A comparative study of the kinetics in spinach thylakoids and in the isolated D1/D2/cytochrome b-559 complex. *Biochimica et Biophysica Acta (BBA) -. Bioenergetics*, *974*(1), 36–43. PMC https://doi.org/10.1016/S0005-2728(89)80163-X

Morris, A. L. C. (2024). Nuclear magnetic resonance spectroscopy in drug discovery, In the previous edition, W.F. Reynolds contributed to this chapter. In McCreath, S. B., & Clement, Y. N. (Eds.), *Pharmacognosy* (2nd ed., pp. 649–684). Academic Press. DOI: 10.1016/B978-0-443-18657-8.00004-9

Mortimer, M., & Taylor, P. G. (2007). *Chemical Kinetics and Mechanism*. Royal Society of Chemistry.

Mouhsine, B., Norlöff, M., Ghouilem, J., Sallustrau, A., Taran, F., & Audisio, D. (2024). Platform for multiple isotope labeling via carbon–sulfur bond exchange. *Journal of the American Chemical Society*, *146*(12), 8343–8351. DOI: 10.1021/jacs.3c14106 PMID: 38498972

Murray, P. R. S. (1977). *Principles of Organic Chemistry*. Heinemann.

Murugasenapathi, N. K., & Palanisamy, T. (2024). Electrochemical Surface-Enhanced Raman Spectroscopy (EC-SERS): Techniques, Applications, and Future Perspectives. In Singh, D. K., Kumar Mishra, A., & Materny, A. (Eds.), *Raman Spectroscopy: Advances and Applications* (pp. 199–214). Springer Nature Singapore. DOI: 10.1007/978-981-97-1703-3_9

Norman, R. O., & Coxon, J. M. (1993). *Principles of Organic Synthesis* (3rd ed.). CRC Press. DOI: 10.1007/978-94-011-2166-8

Nyagilo, V. O., Mallojjala, S. C., & Hirschi, J. S. (2024). Transition state analysis of key steps in dual photoredox-cobalt-catalyzed elimination of alkyl bromides. *ACS Catalysis*, *14*(7), 4683–4689. DOI: 10.1021/acscatal.3c06324 PMID: 39211423

O'Connell, J. F., & Rapoport, H. (1992). 1- Benzenesulfonyl-and 1-p-toluenesulfonyl-3- methylimidazolium triflates: Efficient reagents for the preparation of arylsulfonamides and arylsulfonates. *The Journal of Organic Chemistry*, *57*(17). Advance online publication. DOI: 10.1021/jo00043a046

Ogo, S., Yatabe, T., Miyazawa, K., Hashimoto, Y., Takahashi, C., Nakai, H., & Shiota, Y. (2024). Cyclopropanation using electrons derived from hydrogen: Reaction of alkenes and hydrogen without hydrogenation. *JACS Au*, *4*(4), 1615–1622. DOI: 10.1021/jacsau.4c00098 PMID: 38665645

Ookubo, Y., & Osuka, A. (2012). Synthesis of a library of fluorescent 2-aryl-3-trifluoromethylnaphthofurans from naphthols by using a sequential Pummerer annulation/cross-coupling strategy and their photophysical properties. *Chemistry (Weinheim an der Bergstrasse, Germany)*, *18*(40), 12690–12697. DOI: 10.1002/chem.201201930

Osifová, Z., Socha, O., & Dračínský, M. (2024). NMR-Challenge.com: Exploring the most common mistakes in NMR assignments. *Journal of Chemical Education*, *101*(6), 2561–2569. DOI: 10.1021/acs.jchemed.4c00092

Özbek, N., Katırcıoğlu, H., Karacan, N., & Baykal, T. (2007). Synthesis, characterization and antimicrobial activity of new aliphatic sulfonamide. *Bioorganic & Medicinal Chemistry*, *15*(15), 5105–5109. DOI: 10.1016/j.bmc.2007.05.037 PMID: 17544281

P, N. R. (2011). *Fundamentals of Reaction Mechanisms in Organic Chemistry*. PHI Learning Pvt. Ltd.

Patai, S., Rappoport, H., & Sterling, J. J. (1988). *The chemistry of sulfones and sulphoxides*. Wiley.

Pecoraro, V. L. (Ed.). (1992). *Manganese redox enzymes*. VCH Publishers.

Proniewicz, E., & Lombardi, J. R. (2021). Influence of excitation wavelength, solution pH, incubation time, absolute configuration (S and R) and constitutional isomerism (ortho-, meta- and para-) on the adsorption on the surface of Ag nanoparticles of N-benzylamino(boronphenyl) methylphosphonic ac. *Applied Surface Science*, *563*(May), 150341. DOI: 10.1016/j.apsusc.2021.150341

Pulis, A. P., & Procter, D. J. (2016). C–H coupling reactions directed by sulfoxides: Teaching an old functional group new tricks. *Angewandte Chemie International Edition*, *55*(33), 9842–9860. DOI: 10.1002/anie.201512854

Robertson, H. C. (2018). *On the Reactions of Alkyl Halides with Sodium Phenolate*. Palala Press.

Roecker, L., & Meyer, T. J. (1987). Hydride transfer in the oxidation of alcohols by [(bpy)2(py)Ru(Q)]2+. A kH/kD kinetic isotope effect of 50. *Journal of the American Chemical Society*, *109*(3), 746–754. 10.1021/ja00237a019

Roncone, A., Kelly, S. D., Giannioti, Z., Hauk, C., Caillet, C., Newton, P. N., Perez-Mon, C., & Bontempo, L. (2024). Stable isotope ratio analysis: An emerging tool to trace the origin of falsified medicines. *Trends in Analytical Chemistry*, *174*, 117666. DOI: 10.1016/j.trac.2024.117666

Roos, G., & Roos, C. (2015). Isomers and Stereochemistry. *Organic Chemistry Concepts*, 43–54. DOI: 10.1016/B978-0-12-801699-2.00003-1

Sancéau, J. Y., Maltais, R., Zhou, M., Lin, S. X., & Poirier, D. (2024). Synthesis and characterization of targeted 17β-hydroxysteroid dehydrogenase type 7 inhibitors. *The Journal of Steroid Biochemistry and Molecular Biology*, *242*, 106544. DOI: 10.1016/j.jsbmb.2024.106544 PMID: 38754521

Santos, L. S. (2010). *Reactive intermediates: MS Investigations in Solution*. Wiley-VCH.

Sastre, S., Manta, B., Semelak, J. A., Estrin, D., Trujillo, M., Radi, R., & Zeida, A. (2024). Catalytic mechanism of mycobacterium tuberculosis methionine sulfoxide reductase A. *Biochemistry*, *63*(4), 533–544. DOI: 10.1021/acs.biochem.3c00504 PMID: 38286790

Scozzafava, A., Owa, T., Mastrolorenzo, A., & Supuran, C. T. (2003). Anticancer and antiviral sulfonamides. *Current Medicinal Chemistry*, *10*(11), 925–953. DOI: 10.2174/0929867033457647 PMID: 12678681

Shafir, A. (2016). The emergence of sulfoxide and iodonio-based redox arylation as a synthetic tool. *Tetrahedron Letters*, *57*(24), 2673–2682. DOI: 10.1016/j.tetlet.2016.04.036

Shafirovich, V. Y., Courtney, S. H., Ya, N. Q., & Geacintov, N. E. (1995). Reactions of superoxide radical anion with aromatic compounds in aprotic media. *Journal of the American Chemical Society*, *117*(17), 4920–4929. DOI: 10.1021/ja00121a016

Shamsipur, M., Shafiee-Dastgerdi, L., Talebpour, Z., & Haghgoo, S. (2007). 19F NMR as a powerful technique for the assay of anti-psychotic drug haloperidol in human serum and pharmaceutical formulations. *Journal of Pharmaceutical and Biomedical Analysis*, *43*(3), 1116–1121. DOI: 10.1016/j.jpba.2006.09.038 PMID: 17084055

Shang, L., & Zhang, L., (2017). Redox-neutral α-arylation of alkyl nitriles with aryl sulfoxides: A rapid electrophilic rearrangement. *Journal of the American Chemical Society*, *139*(12), 4211–4217. DOI: 10.1021/jacs.7b01204

Sheradsky, T., & Salemnick, G. (1971). The benzidine-like rearrangement of N,O-diarylhydroxylamines. *Tetrahedron Letters*, *12*(7), 645–648. DOI: 10.1016/S0040-4039(01)96519-0

Silberberg, M. S., & Amateis, P. (2018). *Chemistry: The Molecular Nature of Matter and Change* (8th ed.). McGraw-Hill Education. DOI: 10.1036/0078747872

Silva, A., Mourão, J., & Vale, N. (2023). A review of the lidocaine in the perioperative period. *Journal of Personalized Medicine*, *13*(12), 1699. DOI: 10.3390/jpm13121699 PMID: 38138926

Singh, J., & Yadav, L. D. S. (2010). *Organic chemistry.* Revised Edition. Pragati Prakashan Publisher.

Smith, M. B., & March, J. (2006). *March's Advanced Organic Chemistry* (4th ed.). John Wiley & Sons, Inc. DOI: 10.1002/0470084960

Solomons., Fryhle, C. B., & Snyder, S. A. (2022). *Organic Chemistry*. John Wiley & Sons.

Song, Y., Feng, C., Zhou, D., Ma, Z., He, L., Zhang, C., Yu, G., Zhao, Y., Yang, S., & Xing, X. (2024). Constructing efficient bacterial cell factories to enable one-carbon utilization based on quantitative biology: A review. *Quantitative Biology*, *12*(1), 1–14. DOI: 10.1002/qub2.31

Steigman, J. (1937). *The Walden Inversion and the Reactions of Alkyl Halides with Hydroxylic Solvents*. Columbia University.

Stranix, B. R., Lavallée, J.-F., Sévigny, G., Yelle, J., Perron, V., LeBerre, N., Herbart, D., & Wu, J. J. (2006). Lysine sulfonamides as novel HIV-protease inhibitors: Nε-acyl aromatic α-amino acids. *Bioorganic & Medicinal Chemistry Letters*, *16*(13), 3459–3462. DOI: 10.1016/j.bmcl.2006.04.011 PMID: 16644213

Su, X. Z., Zhang, L. F., Hu, K., An, Y., Zhang, Q. P., Tang, J. W., Yan, B.-C., Li, X.-R., Cai, J., Li, X.-N., Sun, H.-D., Jiang, S.-Y., & Puno, P. T. (2024). Discovery of natural potent HMG-CoA reductase degraders for lowering cholesterol. *Angewandte Chemie International Edition*, *63*(6), e202313859. DOI: 10.1002/anie.202313859 PMID: 38055195

Taiwo, B. J., Miller, A. H., Fielding, A. J., Sarker, S. D., van Heerden, F. R., & Fatokun, A. A. (2024). Ceibinin, a new positional isomer of mangiferin from the inflorescence of Ceiba pentandra (Bombacaceae), elicits similar antioxidant effect but no anti-inflammatory potential compared to mangiferin. *Heliyon*, *10*(1), e23335. DOI: 10.1016/j.heliyon.2023.e23335 PMID: 38332887

Tateishi, Y., McCarty, K. D., Martin, M. V., & Guengerich, F. P. (2024). Oxygen-18 labeling defines a ferric peroxide (compound 0) mechanism in the oxidative deformylation of aldehydes by cytochrome P450 2B4. *ACS Catalysis*, *14*(4), 2388–2394. DOI: 10.1021/acscatal.4c00106 PMID: 38384943

Team, Open University an Introduction to the Chemistry of Carbon Compound (1973). *Substitution Reactions, a N D, Elimination Reactions, a N D, Addition Reactions.*

Thomas, G. (2007). *Medicinal chemistry* (2nd ed.). John Wiley & Sons.

Vajira. P.B. (2013). *Reactive intermediates.*

Venetos, M. C., Elkin, M., Delaney, C., Hartwig, J. F., & Persson, K. A. (2024). Deconvolution and analysis of the ^{1}H NMR spectra of crude reaction mixtures. *Journal of Chemical Information and Modeling*, *64*(8), 3008–3020. DOI: 10.1021/acs.jcim.3c01864 PMID: 38573053

Vincent, J. B., & Christou, G. (1989). Higher Oxidation State Manganese Biomolecules. *Advances in Inorganic Chemistry*, *33*, 197–257.

Walker, M. (2018). *Organic Chemistry 1*. State University of New York Oer Services.

Wermuth, C. G. (2015). Optical Isomerism in Drugs. In *The Practice of Medicinal Chemistry* (4th ed.). Elsevier Ltd., DOI: 10.1016/B978-0-12-417205-0.00018-3

Xia, X., & Gao, Y. (2024). Compound specific, intramolecular, and clumped 13C fractionations in the thermal generation and decomposition of ethane and propane: A DFT and kinetic investigation. *Geochimica et Cosmochimica Acta*, *375*, 50–63. DOI: 10.1016/j.gca.2024.05.001

Yanagi, T., & Yorimitsu, H. (2018). Recent development of ortho-C–H functionalization of aryl sulfoxides through [3,3] sigmatropic rearrangement. *Tetrahedron Letters*, *59*(29), 2951–2959. DOI: 10.1016/j.tetlet.2018.06.011

Yanagi, T., & Yorimitsu, H. (2021). Recent development of biaryl synthesis through sigmatropic rearrangements. *Journal of Synthetic Organic Chemistry, Japan*, *79*(5), 427–438. DOI: 10.5059/yukigoseikyokaishi.79.427

Ye, F., Liao, C., Cheng, H., Zhang, J., Wang, H., Li, Y., Li, H., & Ming, J. (2023). Fundamentals of Organic Chemistry. In *Comprehensive Experiments for Materials Science and Engineering*. DOI: 10.1142/9789811274053_0003

Yiğit, B., Taslimi, P., Barut Celepci, D., Taskin-Tok, T., Yiğit, M., Aygün, M., Özdemir, İ., & Gülçin, İ. (2023). Novel PEPPSI-type N-heterocyclic carbene palladium(II) complexes: Synthesis, characterization, in silico studies and enzyme inhibitory properties against some metabolic enzymes. *Inorganica Chimica Acta, 544*, 121239. DOI: 10.1016/j.ica.2022.121239

Yoshida, S., & Oshima, K. (2009). 2-(2,2,2-Trifluoroethylidene)-1,3-dithiane monoxide as a trifluoromethylketene equivalent. *Organic Letters, 11*(10), 2185–2188. DOI: 10.1021/ol900717d

Yuan, H., Du, Y., Liu, F., Guo, L., Sun, Q., Feng, L., & Gao, H. (2020). Tandem approach to NOBIN analogues from arylhydroxylamines and diaryliodonium salts via [3,3]-sigmatropic rearrangement. *Chemical Communications (Cambridge), 56*(59), 8226–8229. DOI: 10.1039/D0CC02919J PMID: 32555844

Yu, L., Chen, B., Li, Z., Su, Y., Jiang, X., Han, Z., Zhou, Y., Yan, D., Zhu, X., & Dong, R. (2024). Computer-aided automated flow chemical synthesis of polymers. *Giant (Oxford, England), 18*, 100252. DOI: 10.1016/j.giant.2024.100252

Zhang, J.-W., Qi, L.-W., Li, S., Xiang, S.-H., & Tan, B. (2020). Direct construction of NOBINs via domino arylation and sigmatropic rearrangement reactions. *Chinese Journal of Chemistry, 38*(12), 1503–1514. DOI: 10.1002/cjoc.202000358

Zhang, L., & Peng, B. (2019). [3,3]- and [5,5]-Sigmatropic rearrangements of aryl sulfoxides using an 'assembly/deprotonation' technology. *Synlett, 30*(20), 2203–2208. DOI: 10.1055/s-0037-1610257

About the Contributors

Rabab M. Nasser works as Associate. Professor. Researcher at Egyptian Petroleum Research Institute, PhD (Ain Shams University). He has published 26 papers, 2 Books, and 2 Book chapters.

***

Khadra Alomari received her MS in Chemistry from St. Francis Xavier University, Canada, in 2015 and her PhD in Organic Chemistry from York University, UK, in 2022. In 2016 she was hired as a lecturer at Jazan University and is currently an assistant professor, head of the chemistry department, and assistant CEO of Applied College in Samtah. She is interested in organic synthesis and natural products.

Charuwan Thanawiroon is an Assistant Professor at Ubon Ratchathani University in Thailand, specializing in Pharmacy and Pharmaceutical Chemistry. She earned her PhD in Medicinal and Natural Product Chemistry from the University of Iowa, USA. Dr. Thanawiroon has dedicated her career to advancing the field of pharmaceutical sciences, focusing on the analysis of synthetic drugs, medicinal compounds, and pharmaceutical care. Her expertise significantly enhances both research and education in pharmaceutical chemistry, establishing her as a key figure in her academic community.

Bancha Yingngam is working as Associate Professor (Pharmaceutical Sciences and Technology) in the Faculty of Pharmaceutical Sciences at Ubon Ratchathani University, Ubon Ratchathani, Thailand. He received his B.Pharm. and Ph.D. (Pharmaceutical Sciences) at the Ubon Ratchathani, and postdoctoral fellowship at Institute of Pharmaceutical Sciences, University of Graz, Graz, Austria. His areas of interest include artificial intelligence, optimization methods, green chemistry, natural product chemistry, cosmeceuticals, and pharmaceutical nanotechnology. He has authored and co-authored many scientific papers published in international journals, and proceedings of international and national scientific conferences. He has also authored more than 30 book chapters and edited three textbooks.

Index

A

B

C

D

E

F

G

H

I

K

M

www.ingramcontent.com/pod-product-compliance
Lightning Source LLC
LaVergne TN
LVHW081331021026
849161LV00008BA/1930